Fluorescence microscopy is used for studying the distribution of substances which are present in very small amounts. Its techniques are used mainly in biology and medicine, but are also valuable in coal petrology and elsewhere. The best-known application is in immunofluorescence. The high sensitivity of the method makes it ideal for studying the distribution of substances in living cells.

Among the many important topics covered in this book are microfluorometry (measurement of the amount of fluorescence) and microspectrofluorometry (determination of spectra of fluorescence). There are also chapters or sections on the photobleaching of fluorescent substances under irradiation, polarized fluorescence, phosphorescence, time resolution, histochemical studies of enzymes by microfluorometry, and scanning fluorescence microscopy and confocal fluorescence microscopy. Also included is a chapter on flow cytometry by Dr Hans Tanke.

Quantitative fluorescence microscopy

Quantitative fluorescence microscopy

F.W.D. ROST

Professor of Anatomy, University of New South Wales

The right of the
University of Cambridge
to print and sell
all manner of books
was granted by
Henry VIII in 1534.
The University has printed
and published continuously
since 1584.

CAMBRIDGE UNIVERSITY PRESS

Cambridge,
New York, Port Chester, Melbourne, Sydney

Published by the Press Syndicate of the University of Cambridge
The Pitt Building, Trumpington Street, Cambridge CB2 1RP
40 West 20th Street, New York, NY 10011-4211, USA
10 Stamford Road, Oakleigh, Melbourne, Victoria 3166, Australia

First published 1991
Printed in Great Britain at the University Press, Cambridge

British Library cataloguing in publication data
Rost, F.W.D.
Quantitative fluorescence microscopy.
1. Fluorescence microscopy
I. Title
502.823

Library of Congress cataloguing in publication data
Rost, F.W.D.
Quantitative fluorescence microscopy / F.W.D. Rost.
p.   cm.
Includes bibliographical references and index.
ISBN 0 521 39422 8
1. Fluorescence microscopy – Technique.   I. Title.
QH212.F55R67   1990
578′.4—dc20   90-15056   CIP

ISBN 0 521 39422 8 hardback

Contents

Preface

This book is one of a pair of books dealing with fluorescence microscopy. The companion volumes (*Fluorescence microscopy*; Rost, 1991a,b) deal with fluorescence microscopy in general and include a brief overview of quantitative fluorescence microscopy, with emphasis on those aspects which are important to an understanding of fluorescence microscopy. The present volume deals with quantitative fluorescence microscopy in much greater detail. Scanning microscopy is dealt with in this volume because scanning usually involves measurements of fluorescence intensity.

My goals in preparing this book have been to provide, firstly, a detailed description of the structure and use of microfluorometric apparatus; secondly, a description of the basic physical and chemical principles involved; and thirdly, a broad overview of the various applications of quantitative fluorescence microscopy, to alert users to techniques which might be useful to them, and to give references to facilitate obtaining more information on any topic. As in the companion volume, I have assumed that the reader consulting this book will usually have a practical problem and desire a pragmatic answer. Accordingly, the main part of the text describes current techniques and applications; most of the history of the development of quantitative fluorescence microscopy and of its applications has been placed separately in Chapter 16.

For starting me off on the way to these books, I have to thank Professor A. G. Everson Pearse, of the Royal Postgraduate Medical School, London. I had the great privilege of working under him for almost a decade (1965–1974), and it was in his laboratory that I became involved in microspectrofluorometry.

I am very much indebted to those of my friends and colleagues who kindly read and commented on chapters or sections of the book: (in alphabetical order) Associate Professor C.G. dos Remedios, Dr E. Kohen, Mr W. Loeb, Mr R.J. Oldfield, Dr G.L. Paul, Professor P.J. Stoward, Dr A.A. Thaer, and Dr N.G.M. Wreford. Of course, I alone must accept responsibility for all errors of commission and omission.

I am very grateful to numerous friends and colleagues who have given me the hospitality of their laboratories and provided much valuable information. I particularly wish to thank (in alphabetical order): Professor G. Bottiroli and his colleagues of the University of Pavia; Professor T. Caspersson and Dr Martin Ritzén of the Karolinska Institute, Stockholm; Mr Karl-Heinz Hormel of E. Leitz GmbH, Wetzlar, formerly of E. Leitz (Instruments) Ltd; Professor M. van der Ploeg and Professor J.S. Ploem of the University of Leiden; Dr Andreas Thaer, of Helmut Hund Kg, Wetzlar, formerly of E. Leitz GmbH and of the Battelle Institut, Frankfurt; Dr F. Walter of E. Leitz GmbH, Wetzlar; and Mr H. Wasmund, formerly of E. Leitz GmbH, Wetzlar.

Preparation of the book was commenced while I was employed at the Royal Postgraduate Medical School, London, and completed at the University of New South Wales. I am indebted to Library staff of the University for much assistance, and particularly to Mr Andrew Holmick for searches of the literature. Mr Patrick de Permentier, Ms Jenny Flux and Ms Stacey McClelland provided technical assistance and help with bibliographic work. A substantial amount of the text was typed by Mrs Lorraine Brooks. For conversion of word-processor text from an older system to WORD on a Macintosh, I am very grateful to the late Mr Peter Hughes of Macquarie University and to Mr Paul Halasz. I am also indebted to Mrs Gillian Rankin of Macquarie University and to Ms Alicia Fritchle for redrawing diagrams, to Mr Collin Yeo for assistance in photography, and to Mrs Mary Armstrong for secretarial assistance.

Last but not least I wish to express my appreciation of the collaboration of Dr Alan Crowden and Mrs Sandi Irvine of CUP.

Fred Rost
Sydney, 1990

Abbreviations and symbols

a	area of measured field
A	(1) optical absorption
	(2) cross-sectional area of light bundle at the prism
AC	alternating current
A-D	analogue-to-digital
ANS	1-anilinonaphthalene-8-sulphonate
AO	Acridine Orange
atm	atmosphere (1 atm $\approx 10^5$ Pa)
B	background photon count
b	constant
BAO	bis-aminophenyl-oxdiazole
BCECF	2'-7'-bis-(2-carboxymethyl)-5- (and -6-) carboxyfluorescein
BMT	bone marrow transplant
BrdU	5-bromodeoxyuridine
c	concentration of fluorophore
CCD	charge-coupled device
CCTV	closed circuit television
CD	compact disc
CPM	3-(4-maleimidylphenyl)-7-diethylamino-4-methylcoumarin
CRO	cathode ray oscilloscope
d	optical path length (depth)
DANS	dansyl chloride
DAPI	4,6'-diamidino-2-phenylindole dihydrochloride
DASPMI	dimethyl-aminostyryl-methylypyridinium iodide
DASS	defined-substrate sphere system
DC	direct current
DIC	differential interference-contrast
Di-I-LDL	3,3-dioctadecylindocarbocyanine-labelled low-density lipoprotein
DIN	deutsches Industrie Norm
DIPI	4,6-bis-(2-imidazolinyl-4,5H)-2-phenylindole

DNA	deoxyribonucleic acid
DPNH	nicotinamide adenine dinucleotide, reduced form
DTE	dithioerythritol
DTT	dithiothreitol
e	base of natural logarithms (2·71828 . . .)
E	(1) optical extinction (optical density)
	(2) quantum energy of photon
	(3) radiant energy at entrance slit per wavelength unit
EHT	high-voltage supply for photomultiplier tube (extra-high-voltage)
ELISA	enzyme-linked immunosorbent assay
Em	emission maximum
Ex	excitation maximum
eV	electron volts
F	relative fluorescence intensity
f	focal length (of collimator)
FDA	fluorescein diacetate
F-DIM	fluorescence digital imaging microscopy
FIF	formaldehyde-induced fluorescence
FITC	fluorescein isothiocyanate
FPD	fluorescence photoactivation and dissipation
FPR	fluorescence photobleaching recovery
FRAP	fluorescence redistribution after photobleaching
FRP	final reaction product
GVHD	graft versus host disease
h	Planck constant
5-HT	5-hydroxytryptamine (serotonin)
HPD	haematoporphyrin
Hz	Hertz (cycles per second)
I_0	intensity of incident light
I_a	intensity of absorbed light
I_f	intensity of fluorescence
I_r	intensity of reflected light
I_t	intensity of transmitted light
IR	infrared
J	joules
K	(degrees) Kelvin
k	Boltzmann constant
m	metres
MDy	microdensitometry
MFy	microfluorometry
mm	millimetres
MPV	Mikrophotometer mit variable Messblende
MRC	Medical Research Council

ms	milliseconds (10^{-3} s)
MSA	Microscope Spectrum Analyser
MSF	microspectrofluorometer
MSFy	microspectrofluorometry
N	(1) observed photon count
	(2) total photon count
NA	(1) numerical aperture
	(2) noradrenaline
NADH	nicotinamide adenine dinucleotide, reduced form
n_D	refractive index at the sodium D line
n_E	refractive index at the iron E line
nm	nanometres (10^{-9} m)
ns	nanoseconds (10^{-9} s)
OPT	o-phthalaldehyde
p	time
PAS	periodic acid-Schiff
PC	personal computer
P-Con A	pyrene-concavalin A
PMT	photomultiplier tube
PRP	primary reaction product
ps	picoseconds (10^{-12} s)
PVP	polyvinylpyrrolidone
QVIM	quantitative video intensification microscopy
R	percentage polarization
r	a constant
RNA	ribonucleic acid
RNase	ribonuclease
s	second
s	length of capillary tube
S	signal
S_0	singlet ground state of molecule
S_1, S_2	1st, 2nd, excited singlet states of molecule
s.d.	standard deviation
SFM	scanning fluorescence microscopy
SLR	single lens reflex
S_w	slit width
S_h	slit height
T	(1) optical transmission
	(2) absolute temperature
T_0	triplet ground state of molecule
T_1	1st excited triplet state
TMB	tetramethylbenzidine
TPNH	nicotinamide adenine dinucleotide phosphate, reduced form
TRIC	tetramethylrhodamine isothiocyanate

TSRLM		tandem scanning confocal fluorescence microscope
UV		ultraviolet
V		volts
V_{max}		maximum rate of enzyme reaction
VFM		video fluorescence microscopy
VIM		video intensification microscopy
W		watts
W		(1) radiant energy per unit time
		(2) mass of fluorochrome
WL		wavelength
x		amount of unknown substance

Greek letters

Γ	(gamma)	path difference
Δ_n	(delta n)	difference in refractive index
ϵ	(epsilon)	molar extinction coefficient of fluorophore
λ	(lambda)	wavelength
μ	(mu)	refractive index
μm		micrometres (10^{-6} m)
μs		microseconds (10^{-6} s)
μW		microwatts (10^{-6} W)
π	(pi)	ratio of circumference to diameter of circle ($3 \cdot 14159 \ldots$)
ν	(nu)	frequency
σ	(sigma)	standard deviation
τ	(tau)	lifetime of fluorescence
τ_E		lifetime of excited state
τ_F		lifetime of fluorescence
ϕ	(phi)	quantum efficiency

1

Principles of quantitative fluorescence microscopy

Quantitative fluorescence microscopy, as the name implies, is concerned with making measurements from fluorescent specimens in a fluorescence microscope, by measuring fluorescence emission from a defined area or areas of a specimen. The basic measuring technique is called microfluorometry. Quantitative fluorescence microscopy provides part of a range of techniques for microscope photometry; the other microphotometric techniques involve measurement of transmitted or reflected light.

The information to be gained by microfluorometry is of several types. First, microfluorometry is most commonly applied to determining the amount of some specific substance, such as deoxyribonucleic acid (DNA), present in particular regions, such as cell nuclei, by comparison of the intensity of the fluorescence of the specified regions with that of a standard. Secondly, determination of the fluorescence spectra, and possibly of other characteristics described below, may enable the identification of specific fluorescent substances. Thirdly, determination of parameters of fluorescence of a known fluorophore introduced intracellularly as a probe can give information about the micro-environment of the probe, and thereby about the inside of the cell. Fourthly, scanning devices can quantify the distribution of fluorescent components in tissue or other material, and enable digitized images to be built up which can be examined, recorded, corrected, and subjected to image analysis.

Probably the most exciting recent developments in fluorescence microscopy have been the commercial availability of confocal fluorescence microscopy (described in Chapter 15), which enables three-dimensional analysis by optical sectioning, and the development of video-intensified fluorescence microscopy, which allows living cells to be studied with minimal radiation.

Fluorescence microscopy

Fluorescence microscopy is described in the companion volumes (Rost, 1991a,b). A brief summary of the nature of fluorescence microscopy, to remind the reader (if necessary) of the major features, is given in this section.

In a fluorescence microscope, the specimen is illuminated with light of a short wavelength, e.g. ultraviolet (UV) or blue. Part of this light is absorbed by the specimen, and re-emitted as fluorescence. The re-emitted light has a longer wavelength than that of the incident light. To enable the comparatively weak fluorescence to be seen, despite the strong illumination, the light used for excitation is filtered out by a secondary (barrier) filter placed between the specimen and the eye. This filter, in principle, should be fully opaque at the wavelength used for excitation, and fully transparent at longer wavelengths so as to transmit the fluorescence. The fluorescent object is therefore seen as a bright image against a dark background.

It follows that a fluorescence microscope differs from a microscope used for

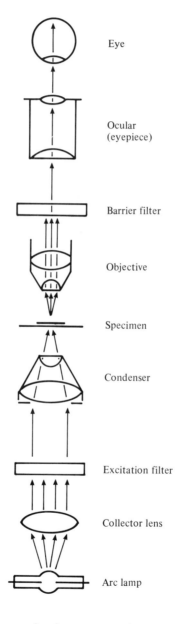

Fig. 1.1. Optical diagram of a fluorescence microscope using dia-illumination (illumination through the specimen) with brightfield substage condenser.

conventional absorption microscopy mainly in that it has a special light source and a pair of complementary filters. The basic arrangement is shown in Fig. 1.1. The lamp should be a powerful light source, rich in short wavelengths: high-pressure mercury arc lamps are the most common. A primary (excitation) filter is placed somewhere between the lamp and the specimen. This filter, in combination with the lamp, should provide light over a comparatively narrow band of wavelengths corresponding to the absorption maximum of the fluorescent substance (fluorophore). The secondary (barrier or suppression) filter prevents the excitation light from reaching the observer's eye (or a photometric device in place of the eye), and is placed anywhere between the specimen and the eye, preferably in the body tube. Its transmission should be as low as possible in the spectral range of the light used for excitation, and as high as possible within the spectral range of the emission from the specimen.

The functions of the excitation and barrier filters may be more clearly demonstrated with the aid of Fig. 1.2. The figure relates to a hypothetical fluorophore (fluorescent object) which absorbs in the blue and fluoresces in the green. The absorption spectrum of the fluorophore is shown on the left; this indicates that the substance absorbs predominantly in the blue, hardly at all in the violet, and the spectrum has a long tail with a secondary peak in the UV. The emission curve is an approximate mirror image of the longest-wavelength peak of the excitation curve.

The broken lines show ideal curves for the transmission of the two filters. The excitation filter should ideally have complete (100%) transmission in the region of the absorption peak of the fluorophore, while having zero (0%) transmission (complete opacity) at the wavelengths of fluorescence. Conversely, the barrier filter should have 100% transmission at the wavelengths of fluorescence, with 0% transmission in the region used for excitation. The two filters are therefore complementary; they allow the

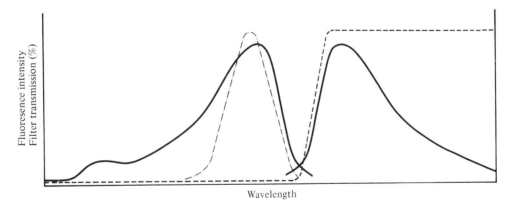

Fig. 1.2. Excitation and barrier filters in relation to the spectra of a fluorophore. Solid lines (–), the excitation (left) and emission (right) spectra of a hypothetical fluorophore. This fluorophore has maximal absorption in the blue, with a smaller absorption peak in the UV; the emission spectrum has a peak in the green, with a long tail extending into the red. The transmission of an excitation filter (-·-·-·-) with a narrow bandpass in the blue provides optimal excitation. A barrier filter (-----) is opaque at the excitation wavelengths, while transmitting the major part of the emission.

passage of short-wavelength light as far as the specimen, blocking its passage to the eye, and transmitting the fluorescence. For quantitative studies, one or both of the filters is commonly replaced by a monochromator, a device which transmits light in a selected narrow band of wavelengths only.

Opaque or very thick objects can be examined using epi-illumination, a technique whereby the light for excitation is reflected downwards through or around the objective onto the specimen. Although this technique is in principle only essential for examination of opaque objects, such as the surface of intact organs, it has some advantages also for the more conventional transparent specimens, and particularly for quantification, as will be explained below.

Because the image seen in the microscope may consist of only a few small fluorescent areas in an otherwise black field, fluorescence microscopy is sometimes supplemented with other forms of microscopy, e.g. phase-contrast, to enable the specimen as a whole to be visualized and to show the position of fluorescent areas in relation to the rest of the specimen.

Fluorescence microscopy has two particular problems, apart from the purely technical one of having the necessary equipment. The fluorescence image as seen in the microscope is weak compared to that obtained by almost all other kinds of microscopy. This makes particular demands on the efficiency of the system, to avoid loss of light. The dimness of the image may lead to difficulties in interpretation, due to the eye's poor discrimination of colour at low light levels. To make matters worse, the specimen usually fades more or less rapidly under irradiation, producing errors during photometry, and may fade too quickly to be photographed.

As in other forms of microscopy, there are three basic kinds of fluorescence microscopy: qualitative, quantitative and analytical. The first is concerned with morphology, or with whether or not something (e.g. an immunological reaction) is present. Quantitative fluorescence microscopy is concerned with finding out how much of a specific substance is present in a specified region of the specimen. Analytical fluorescence microscopy is the characterization of a fluorophore by measurement of excitation and emission spectra or other characteristics such as polarization or decay time.

Quantification

Microfluorometry basically involves measuring the brightness of fluorescence emission from a defined area of the specimen under standardized conditions. This process can be extended to the measurement of other characteristics of the fluorescence: the actual experimental variables which can be measured or determined are the fluorescence intensity, excitation and emission wavelengths, the polarization of fluorescence, the time, and the area or volume from which measurement is made. It may be possible to measure more than one variable simultaneously (e.g. see Araki & Yamada, 1986). In principle, microdensitometric measurements can be made from the same specimen (given suitable apparatus, see Rost & Pearse, 1971) and the results combined with microfluorometric measurements.

The measuring field is a region of the field delineated by a field stop (diaphragm) in the magnifying system, within which measurement is made. The diaphragm itself may be either a variable circular iris diaphragm, variable rectangular diaphragm, or one of a set of interchangeable fixed diaphragms (e.g. holes in a metal plate).

The parameters or characteristics of the fluorescence which can be measured from a given region of the specimen are as follows.

1. The intensity of fluorescence (i.e. the emittance, I_f).
2. The emission spectrum.
3. The excitation spectrum.
4. The quantum efficiency (ϕ) of the fluorophore.
5. The polarization of fluorescence.
6. The fluorescent lifetime (v) of the fluorophore.
7. Structure-correlated information, e.g. the relative areas of fluorescence and non-fluorescence.

Changes in all of the above can be followed over a period of time, either short (comparable to the fluorescent lifetime) or long (e.g. rate of fading). In addition, information may be obtainable concerning the transfer of energy between different probes.

Of the above possible types of measurement: (1) is microfluorometry proper, and is the main subject of the present chapter and Chapters 2–4; (2) and (3) are microspectro-fluorometry, and are dealt with in Chapters 6–10; (5) and (6) have only recently been applied to fluorescence microscopy, see Chapter 12; (7) requires scanning, either mechanically or with a video camera, and is dealt with in Chapter 15; (8), kinetic measurement, is dealt with in Chapters 11–15. The measurement of cells *en passant* in a flow system (flow cytofluorometry) is dealt with in Chapter 14.

The standard text on microscope photometry in general is that of Piller (1977). Many topics related to quantitative cytochemistry are discussed in detail in the books edited by Kohen & Hirschberg (1989), Thaer & Sernetz (1973), Wied & Bahr (1970), and Wied (1966), and in the journal *Cytometry*. Brief reviews of microfluorometry were given by Ploem & Tanke (1987), Rost (1980, 1974), Fukada, Böhm & Fujita (1978), Ploem (1977), and Ruch & Leeman (1973). Useful information is also found in the chemical and biochemical literature on fluorometry: see, for example, Miller (1984), Wehry (1982), O'Donnell & Solie (1976), Wotherspoon, Oster & Oster (1972), and Parker (1969a).

Microfluorometry

For the remainder of this chapter, and Chapters 2–5, microfluorometry may be said to be a technique whereby the intensity of fluorescence is measured from a given area of a specimen, usually with a view to measuring the amount of fluorophore present and thereby estimating the amount of some substance present in that region of the specimen. For example, nuclear DNA may be measured by microfluorometry of nuclei subjected to Feulgen hydrolysis and stained with a fluorescent Schiff-type reagent. Further examples are described in Chapter 5.

    A microfluorometer, the instrument required for microfluorometry, is essentially a fluorescence microscope with a measuring device which measures the intensity of fluorescence from a specified area of the specimen (see Fig. 1.3). This measured intensity is compared to the intensity to be measured from a standard containing a known amount of the fluorophore, and the amount of fluorophore present in the specimen is determined by proportion.

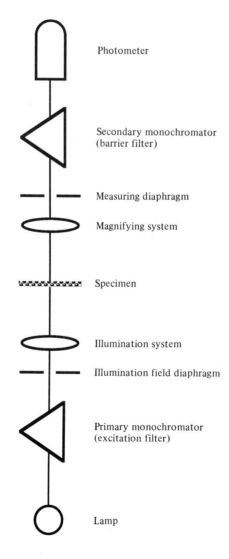

Photometer

Secondary monochromator
(barrier filter)

Measuring diaphragm

Magnifying system

Specimen

Illumination system

Illumination field diaphragm

Primary monochromator
(excitation filter)

Lamp

Fig. 1.3. Block diagram of a microfluorometer. See also figures in Chapter 2.

Although simple in principle, microfluorometry presents a number of traps for the unwary, against which due precautions must be taken; these are discussed in Chapter 4.

Sources of error

All microfluorometry is based on the concept of irradiating a specimen with nominally monochromatic light of known intensity, and measuring the intensity of light emitted from a given area of the specimen. It is generally assumed that the intensity of fluorescence is proportional to the amount of fluorophore present. Unfortunately this is only true at low concentrations of fluorophore, as is shown in Chapter 4. If the excitation (absorption) and emission curves of the fluorophore overlap, as they commonly do, some of the emitted light may be reabsorbed (see Chapter 4). For this reason, it is important that the fluorescence emission be measured at a wavelength long enough to be beyond the effective limits of the absorption spectrum of the fluorescent substance. A further difficulty is fading of the specimen as a result of irradiation during measurement, and/or during initial examination to find the area to be measured. Causes of errors are discussed in Chapter 4.

Standardization

In microfluorometry, the observable data are the fluorescence intensity of the specimen and of a standard containing a known amount of the fluorophore, and the ratio of these measurements is calculated. The incident light and the emission are at different wavelengths and radiate in different directions (emission radiates from the specimen in all directions), so that they cannot be compared directly. This contrasts with the situation in densitometry, where incident and transmitted light are measured with the same apparatus, respectively in an empty field and with the specimen present. Accordingly, quantification requires comparison with a standard, measured under the same conditions as those of the specimen. Standards are discussed in Chapter 4.

Instrumentation

Practical instrumentation is the subject of the next chapter. However, the basic principles involved need to be set forth here. A microfluorometer is essentially a fluorescence microscope with a stable, uniform light source and a photometric device for measuring the intensity of fluorescence from a defined area of the specimen (see Fig. 1.3). Usually, a separate illumination system is required for phase-contrast or other examination with light of a longer wavelength, so that the specimen can be set in the measuring field before the excitation light is allowed to irradiate the specimen; this procedure reduces fading.

The direction of illumination

It was Rigler (1966) who first pointed out the significance of the relative directions of the illumination and of the measurement of fluorescence emission in relation to the accuracy of the measurement. He concluded that epi-illumination was to be preferred; see also Benson & Knopp (1984). The reasons are as follows.

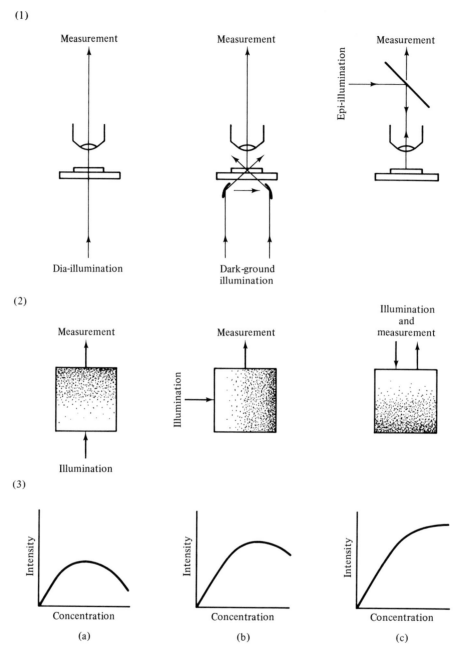

Fig. 1.4. The relationships between the concentration of a fluorophore and the intensity of the measured fluorescence as observed with three optical arrangements. The first row (1) shows optical systems in a microfluorometer: (a) dia-illumination (with a substage condenser), (b) darkground illumination, and (c) epi-illumination. The second row (2) shows the nearest corresponding arrangement with a cuvette in a chemical fluorometer. The third row (3) shows the relationship between concentration of fluorophore (horizontal axis) and measured fluorescence intensity (vertical axis). The advantage of epi-illumination over dia-illumination and darkground illumination is evident. Diagram modified from Rost (1980); after Udenfriend (1962) and Rigler (1966).

1. Error due to reabsorption of fluorescence is reduced (see Fig. 1.4).

2. Correct focussing of the objective automatically focusses the illumination system, thereby maintaining standard conditions from one measurement to the next. Mutual centration of 'condenser' and 'objective' is assured.

3. With high-power objectives such as are usually required for microfluorometry, brighter illumination is obtained.

4. Irradiation is confined to the area being examined.

5. The high optical correction of microscope objectives, far exceeding that of condenser systems, enables high-quality imaging of the field diaphragm into the object plane, homogeneous distribution of light intensity, and reduction of light scatter.

The simultaneous use of phase- or interference-contrast and fluorescence is facilitated by epi-illumination for fluorescence. Independent field diaphragms enable the fluorescence field diaphragm to be kept small, to minimize irradiation of the specimen.

Sensitivity

A high degree of sensitivity is not always necessary (e.g. for determination of nuclear DNA with a fluorochrome), but may be invaluable for the assay of tiny amounts of substances such as neurotransmitter amines. The smallest amount of fluorescent substance which can be assayed in a given specimen depends upon the sensitivity of the microfluorometer, and on the amount of background fluorescence. The sensitivity of the microfluorometer depends on the brilliance of the light source, the efficiency of the optical system, and the sensitivity of the detection system. The amount of background depends in part on the nature of the specimen (e.g. the amount of autofluorescence) and on the amount of stray light in the system.

Since the number of photons emitted is proportional to the number absorbed, the sensitivity of the emission measurement will depend partly on the brightness or radiant density of the exciting light source. Because the image of a light source cannot radiate at a higher density than the source itself, the amount of excitation which can ultimately be achieved is dependent upon the brightness of the lamp. The illumination system produces in the object plane a reduced (demagnified) image of the light source; its brilliance is dependent upon the square of the numerical aperture (NA) of the system (generally determined by the aperture iris diaphragm of the condenser) and upon the square of the demagnification of the image of the lamp (the smaller the image, the brighter, but only to a limited extent; the image can never be more intense than the source). Generally, for microfluorometry, only a relatively small field need be illuminated, so that arc lamps with a small but brilliant (and stable) arc are most suitable. Similarly, the observation system focusses an image of the specimen upon the eye; the brilliance of the image depends upon the square of the NA of the system (determined by the objective) and inversely upon the square of the image magnification (determined jointly by the objective, the ocular, and any intermediate magnifying device).

The brightness of the excitation irradiation is maintained by choosing a high-aperture illumination system, and the most efficient monochromation system. For the latter, filters are usually preferable to a monochromator, because a greater aperture is obtainable at minimal expense. Monochromators are described in Chapter 7.

However, the more brilliant the excitation, the more rapid will be the consequent fading. Therefore, in most cases it is more appropriate to increase sensitivity of the system by increasing the sensitivity of the detector rather than by increasing the brilliance of excitation.

Just as, in the illumination system, greatest brightness is obtained by the combination of a brilliant source and an efficient optical system, so in the measuring system greatest sensitivity is obtained by a combination of an efficient optical system and a sensitive detecting device. Maximization of the efficiency of the optical system is discussed in *Fluorescence microscopy* (Rost, 1991a) as regards the use of high-aperture objectives, and a low-power subsequent magnifying system, and minimization of losses by reflections at air–glass interfaces by using as few optical elements as possible. The barrier filter may take the form of either a conventional long-pass filter, or a band-pass filter (interference filter or monochromator). The former gives greatest sensitivity, the latter may give increased optical specificity, including some measure of freedom from stray light at wavelengths not relevant to the measurement. The sensitivity of the detecting device depends upon the device itself, and on the electronic system used for measurement. At the present time, the most sensitive apparatus generally available appears to be photomultiplier tubes with photon-counting circuitry (see Chapter 2).

Aperture-defined volume

In thin specimens or opaque surfaces, microfluorometric measurements are made from an area defined by a diaphragm in the measuring system. If, using epi-illumination, the objective is dipped into a fluid, or applied to the surface of a translucent solid, the region from which measurement is made is a volume, forming a cone defined by the aperture of the objective. This is referred to as *aperture-defined volume*. This is useful for uranyl glass standards (Jongsma, Hijmans & Ploem, 1971) and Terasaki trays (Deelder, Tanke & Ploem, 1978).

Comparison with microdensitometry

It is natural to compare quantification by microfluorometry with quantification by microdensitometry, particularly as very often both techniques could in principle be used for the same basic purpose, as in the Feulgen procedure. There are two main methods available for microdensitometry: the scanning method and the two-wavelength method. Details are given by Piller (1977), Wied (1966) and Wied & Bahr (1970). The main points of comparison of quantification by microfluorometry and by microdensitometry are summarized in Table 1.1.

Theoretically, a major difference between microfluorometry and microdensitometry is that (generally speaking) microfluorometry does not suffer from distributional error

Table 1.1. **Comparison of microfluorometry and microdensitometry**

Characteristic	Microfluorometry	Microdensitometry
Sensitivity	High	Low
Accuracy at		
Low concentrations	Normal	Low
High concentrations	Low	Normal to $E \approx 0.4$ [a]
Opaque specimens	Possible	Impossible
Optical specificity	At 2 wavelengths (excitation and emission)	At 1 wavelength (absorption)
Decay time resolution	Possible	Impossible
Type of measurement	Relative	Absolute
Speed of measurement	Fast	Slow
Distributional error	Absent [b]	Usual

Notes:
[a] i.e. extinction less than about 0·4.
[b] Except at extremely high concentrations extending above the linear region.

(the exception being in the presence of very high local concentration, when the intensity of fluorescence is no longer proportional to the concentration of the fluorophore). This in turn leads to one of the advantages which microfluorometry has over scanning microdensitometry: speed. Once the required area of the specimen has been selected, a measurement may be made by microfluorometry almost instantly; whereas by scanning microdensitometry, each area to be measured usually requires many point measurements. It must be admitted that absorption measurement by the two-wavelength method can also be made instantaneously, the two measurements at different wavelengths being made simultaneously using a dichromatic mirror and two detectors.

In general, microfluorometry is advantageous for assaying low to very low concentrations, for which it offers greater sensitivity. This is brought out by the data in Table 4.1, which shows that, at very low extinctions, measurement of absorption (by fluorometry) is more accurate than measurement of transmission; whereas at high extinctions, the reverse is the case. In some cases, only one method or the other is available, the histochemical reaction being quantified giving a product either non-fluorescent (microfluorometry not possible) or not absorbing in the visible region (UV microdensitometry may be too costly). Sometimes both methods are available, as in quantification of DNA stained by the Feulgen method with Pararosaniline. For examples of experimental comparison of the two methods, see Böhm, Sprenger & Sandritter (1970), and Smith, Redick & Baron (1983).

Opaque specimens

Because of the possibility of epi-illumination, microfluorometry can be employed on opaque specimens, for which microdensitometry is impossible. Microfluorometry is particularly useful in studying the NADH fluorescence of the surface of intact organs (see Chapter 13). Microspectrofluorometry (in which spectra are measured) can similarly be carried out on opaque specimens, such as polished specimens of coal.

Optical specificity

In microdensitometry, some degree of optical specificity is obtained by employing monochromatic light of a wavelength corresponding to the absorption maximum of the desired substance. In microfluorometry, two wavelengths are involved (excitation and emission) and in principle optical specificity can be maximized by appropriate choice of wavelength for excitation and wavelength band for emission measurements. In addition, the technique automatically differentiates fluorescent from non-fluorescent substances, and the fluorescence decay times can be used to differentiate between different fluorophores (see Chapter 12).

2

Instrumentation for microfluorometry

Microfluorometers, for the purpose of this chapter, are those instruments which operate at one or more fixed wavelengths, using filters for monochromation, possibly with a laser light source. Microspectrofluorometers (defined as possessing at least one variable monochromator) are dealt with in Chapter 7; they can of course be used for the same purposes as the simpler instruments described in this chapter. Equipment for measurements with video and other scanning devices is discussed in Chapter 15.

This chapter is divided into two sections: the first deals with practical realization of equipment, both experimental and commercial, and the second deals with individual components in sequence from the light source to the photometer. Older instruments, and the history of their development, are described in Chapter 16.

MICROFLUOROMETERS

Microfluorometers fall into two practical categories: standard commercial equipment, and instruments constructed by the user from more-or-less standard components. The latter tend to be widely used, usually on the basis that it may be rather less expensive to adapt existing equipment than to buy a complete system, or that if one is assembling a system oneself it may be possible to select the best items from each of a number of rival manufacturers. For example, Tiffe (1977) combined a Zeiss Universal microscope, a Leitz MPV-1 photometer, a Zeiss amplifier, a Siemens digital voltmeter, and a Philips chart recorder. However, there is no doubt that it is simpler to buy a complete system from a single manufacturer, and the range of currently available commercial equipment should satisfy most needs.

Numerous microfluorometers have been described in the literature. The first generation of these employed, as sources of monochromatic light for excitation, arc lamps and filters; the second generation employ lasers. These two basic types will be discussed separately.

ARC AND FILTER MICROFLUOROMETERS

Numerous instruments of this type have been described in the literature (see Chapter 16). However, equipment for microfluorometry has been available commercially for quite some years, and it is no longer necessary for each worker to design and build a new instrument. Leitz, in particular, appear to have paid attention to the development of equipment for microfluorometry. Independent companies also supply some equipment which can be used in conjunction with a microscope.

Leitz

Leitz have made several photometer units over the years, in the modular MPV system (the acronym MPV stands for *Mikrophotometer mit Variable Messblende*, signifying an advance over the earliest model which had fixed field diaphragms only). There are currently (1990) four MPV models: MPV Compact, MPV-3, MPV-SP, and MPV-MT2.

In general, the MPV microscope photometers attach to a standard Leitz microscope, other functions of the microscope (observation, photography) being undisturbed. Measurements are made at an emission wavelength selected by a barrier or interference filter, from an area determined by a measuring field diaphragm incorporated into the photometer unit. Epi-illumination is standard, using a dichromatic beamsplitter. Monochromation of the excitation can be obtained by using interchangeable interference filters (for mercury arc) or by a Leitz grating monochromator, in conjunction with a xenon arc (see Chapter 7). This instrumentation is versatile and many variations are possible with standard components. Stabilized arc sources are available but no provision is made for a reference channel. Microfluorometers based on an earlier model of this photometer were described by Thaer (1966a).

The MPV Compact photometer attachment (Fig. 2.1) can in principle be adapted to any Leitz laboratory microscope. The current (1990) model is the MPV Compact 2. It sits on top of the FSA trinocular body tube, which has a built-in reflecting system that makes possible simultaneous observation of images of the specimen and the measuring field diaphragm: this body tube is used in a similar way for photomicrography with the Orthomat E.

The photometer unit contains four main components: the measuring field diaphragm, a spectral filter, a pilot lamp, and a photomultiplier. The field diaphragm and spectral filter are fitted in interchangeable holders which fit into slots in the photometer housing. The interchangeable measuring field diaphragms are available in three basic varieties: circular with adjustable iris, rectangular with side lengths adjustable independently, and a set of circular fixed diaphragms of various sizes. The holder for the latter can accommodate home-made diaphragms of special size and shape.

The filter slot accommodates interference filters in interchangeable holders: a filter of appropriate wavelength is chosen for each application. Alternatively, a variable interference filter can be used: these are rectangular in shape, and have varying peak transmission wavelength along their length. Hence, by moving the filter past a slit, any desired wavelength can be chosen within the range of the filter. The fixed-wavelength filters are more consistent and convenient for routine use. An empty filter holder can be used if the filters in the epi-illumination block are sufficient and appropriate for photometry.

The pilot light illuminates the measuring field diaphragm from above. During focussing and selection of the desired object for measurement, an image of the measuring field diaphragm illuminated by the pilot lamp is reflected into the binocular observation tubes by a deflecting mirror in the photometer unit and the reflecting system of the FSA tube. The deflecting mirror is automatically swung out during measurement.

For microfluorometry, the MPV Compact can be fitted to any Leitz fluorescence microscope fitted with the FSA trinocular body tube. It is most conveniently fitted to those stands which have integrated epi-illumination systems, namely the Aristoplan, Fluovert and Orthoplan. A stabilized light source is essential. This is usually based on a 100 W mercury arc lamp or 75 W xenon arc lamp, in the lamphouse 100Z with a clear glass (unfrosted) condenser. A stabilized power unit is required. The xenon arc is mainly useful (in this context) for blue-light excitation. On the Orthoplan microscope, the optional Mirror House 500 enables separate sources to be used for epi-illumination observation (of the entire field of view) and measurement, as well as a separate lamp for transmitted-light observation.

The normal epi-illumination system appropriate to the microscope stand is utilized. For photometry, it is usually preferable to use narrow bandpass filters, such as those in blocks E3, I2/3, K3, L3, M2 and N2 in the interest of specificity. On the other hand, the

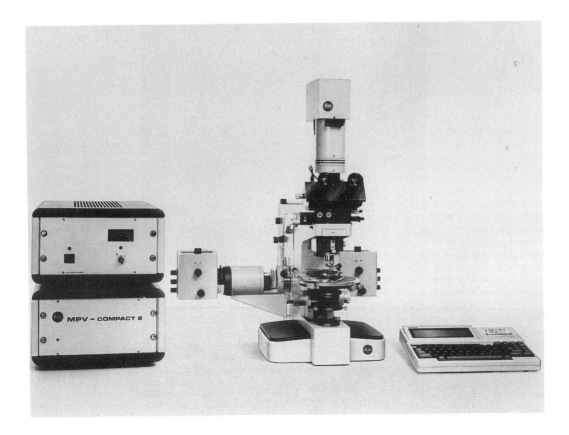

Fig. 2.1. Leitz MPV Compact 2 system on Orthoplan microscope. Note the separate illumination systems for observation and measurement. Photograph courtesy of Dr W.J. Patzelt of E. Leitz GmbH, Wetzlar.

wider-band filters transmit more light and therefore give greater sensitivity. I tend to prefer a narrow-band excitation filter, giving highest specificity and minimal fading, in conjunction with a normal barrier filter for greatest sensitivity of measurement.

The microscope can be fitted with a variety of accessories. Those particularly relevant to microfluorometry are provision for phase-contrast or differential interference-contrast (DIC) examination, an automatic shutter in the excitation pathway, and a motorized scanning stage. A monochromator for excitation can also be added (see Chapter 7).

The MPV-MT2 (Fig. 2.4) is a specialized instrument based on an inverted microscope, and is described below under Inverted microscopes.

The MPV-SP and MPV-3 are more complex photometric systems, designed primarily to incorporate at least one monochromator. They are therefore described with other microspectrofluorometers in Chapter 7.

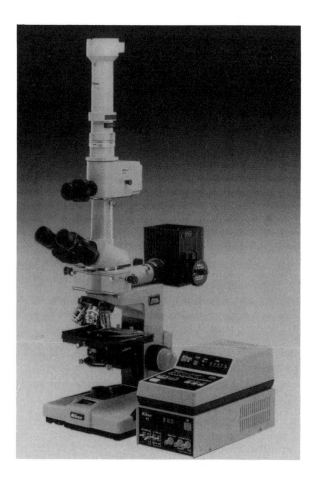

Fig. 2.2. Nikon P1 photometer on Optiphot microscope with fluorescence epi-illumination system. Photograph courtesy of FSE Pty Ltd, Sydney.

Nikon

A photometric module, microphotometry system P1, is available for the larger Nikon microscopes; it is attached to the special port for a photometer or video camera on either an upright microscope (Fig. 2.2) or an inverted microscope (Fig. 2.5; see below).

The Microphot-FXA microscope, equipped with the P1 photometer, has considerable potential for automation. A remote switching unit provides centralized control of automatic options, and there is provision for linking to a computer for fully automatic control. Automatic functions include an autofocus attachment (automatic focussing with low-power objectives, $\times 1$ to $\times 20$), motorized nosepiece and motorized substage condenser. A computer software package, called PHOSCAN-1, is available for IBM and some compatible personal computers to be used with the P1 for operating the photometer, moving a scanning stage, and data acquisition.

Olympus

The BH2-QRFL microfluorometer (Fukuda *et al.*, 1982) was designed for multiparameter cell analysis, using automatic selection of filter sets and data recording. It is based on the Olympus BHS microscope, with a modified fluorescence epi-illumination module incorporating a photometer. A mirror above the epi-illumination system diverts light through a supplementary barrier filter to a photomultiplier. The measuring field is defined by a pinhole in the illumination system; there is no separate measuring diaphragm. A reference channel constantly monitors the excitation intensity; this is achieved by a partially reflecting mirror placed after the barrier filter. A mechanical chopper in the light path is used in conjunction with a lock-in AC amplifier to improve the signal-to noise ratio. For multi-wavelength photometry, combinations of excitation filters, dichromatic mirrors and barrier filters can be computer-controlled. Data processing and control is carried out by a computer via the standard HP-IB interface system. Examples of its use are described by Fukuda (1983). Unfortunately this instrument is available only in Japan.

Zeiss (Oberkochen)

Zeiss have made several microphotometers, and Zeiss microscopes have been the basis of a number of independently built devices.

The MPM 200 (Fig. 2.3) is the current main Zeiss microscope photometer. The MPM 200 on the Axiovert provides photometric possibilities on an inverted microscope. The Axiophot microscope, recently introduced, has a separate port for the attachment of a photometer or video camera. Remote-controlled light shutters provide automatic changes between transmitted light and fluorescence modes, and for short-time excitation. An electromagnetic filter changer is available for dual-wavelength fluorometry, enabling changes in as little as 0·3 s. The photometer unit has auxiliary illumination for centring of the measuring diaphragm, a centrable slot for the measuring diaphragm, a light shutter, and a slot for colour filters. The eyepieces are fitted with a graticule showing concentric circles matched with measuring diaphragms. The binocular tube is provided with an eyepiece shutter to exclude extraneous light. These arrangements are designed to be particularly suitable for dual-wavelength fluorometry, e.g. for measurements of intracellular ions (see Chapter 10).

The MPM 03 with tube head 03 form modules for the UEM and some discontinued microscopes. In this photometer, the measuring field and/or surrounding field and measuring diaphragm are visible simultaneously during observation and measurement. This is achieved by forming an image of the object on a mirror which has perforations acting as stops. During measurement, the full light from the measuring field can be sent to the photomultiplier. Alternatively, an adjustable rectangular diaphragm is available, in a unit which incorporates a semi-transparent mirror to reflect an image of the diaphragm into the observation tube. The spectral range of the basic instrument is determined by the tube head; type 03 limits operation to the visible, type 03 UV extends the range to the UV and IR; a video camera or IR converter can be fitted for observation. The photometric module can be combined with other accessories

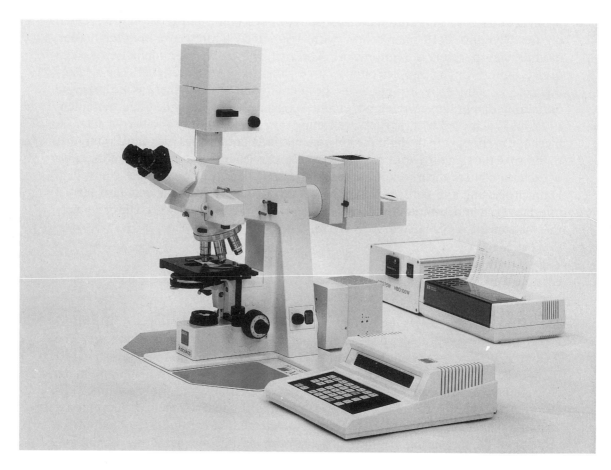

Fig. 2.3. Zeiss MPM 200 photometer system, for microfluorometry on the Axioplan microscope. Photograph courtesy of Carl Zeiss Pty Ltd, Sydney.

as required. Monochromators can be added to convert the instrument for use as a microspectrofluorometer (see Chapter 7).

INVERTED MICROSCOPES

An inverted microscope with an epi-illuminator is particularly suitable for experimental arrangements requiring manipulation of the specimen (Kohen & Legallais, 1965; Kohen, Kohen & Thorell, 1969) or for studying fluid, tissue culture cells and other sediments. Chance & Legallais (1959) were the first to adapt an inverted microscope for microspectrofluorometry. Commercial microfluorometers with inverted microscopes are now available from Leitz, Nikon and Zeiss (see above). Photometric adaptation is particularly easy with the Leitz Fluovert inverted microscope, which has a normal trinocular body tube at the top of the microscope, identical with that of Leitz upright microscopes. The Leitz MPV-MT2 (Fig. 2.4) is a specialized instrument based on an inverted microscope, for analysis of ELISA (enzyme-linked immunosorbent assay) tests and for cell culture studies. As well as standard specimen slides, it accepts Terasaki, microtitre and Hamax trays. A microprocessor controls a scanning stage and filter changing in a motorized epi-illuminator. Semi-automatic and fully automatic measurement procedures can be carried out under microprocessor control.

Fig. 2.5 shows the Nikon Diaphot inverted microscope, with the P1 photometer (referred to above) attached to a port on the side.

LASER MICROFLUOROMETERS

Laser microfluorometers have been described by Kaufman, Nester & Wasserman (1971), Andreoni *et al.* (1975), Prosperi *et al.* (1983), Docchio *et al.* (1984), and van Geel

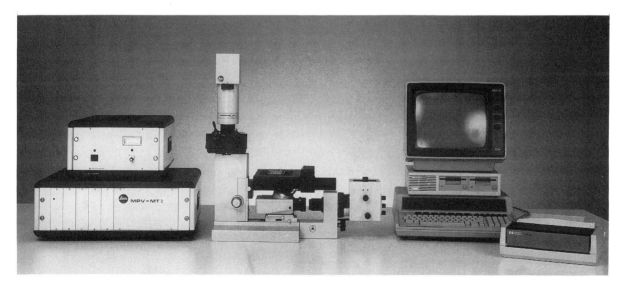

Fig. 2.4. Leitz inverted microscope with fluorescence epi-illuminator and MPV-MT2 photometer. Photograph courtesy of Dr W.J. Patzelt of E. Leitz GmbH, Wetzlar.

et al. (1984). They have the advantage of a monochromatic light source, giving high optical specificity of excitation, and a light source which can be pulsed for brief excitation either to reduce fading or for studies of decay time (for which see Chapter 12). No excitation filter is required, and the requirements for the barrier filter are simplified by the monochromaticity of the source. Selection of a laser system with an appropriate output wavelength enables the excitation wavelength to be matched to the peak of the excitation spectrum of the fluorophore, giving increased specificity and intensity of fluorescence. On the other hand, care must be taken with the optical system to avoid interference effects due to the coherent nature of the source, and laser systems are expensive.

Kaufman *et al.* (1971) described the comparison of two lasers with a 150 W xenon arc for fluorometry. They investigated a helium-cadmium laser with output at 442 nm, and an argon laser with output at 488 nm. Each laser emitted a beam, 3 mm diameter, with continuous powers of 50 μW and 4 μW, respectively. In comparison with the xenon arc, when used for exciting fluorescein isothiocyanate (FITC), the lasers gave stronger fluorescence, due partly to a better match between the excitation peak of the dye and the output wavelength of the lasers, and partly to greater intensity of the source.

Prosperi *et al.* (1983) employed an argon laser with a Leitz MPV2 microfluorometer. Fluorescence emission was selected by dichromatic mirror TK495 and barrier filter K570. The photomultiplier was EMI type 9558, which has extended sensitivity to red;

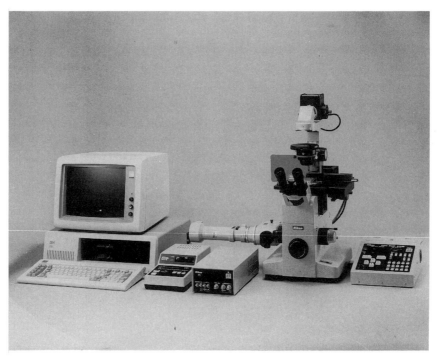

Fig. 2.5. Nikon P1 photometer on Diaphot TMD inverted microscope. Photograph courtesy of FSE Pty Ltd, Sydney.

measurement was by photon counting (Cova, Prenna & Mazzini, 1974). They applied the instrument to the intracellular uptake of anthracycline antibiotics.

Docchio *et al.* (1984) used a pulsed nitrogen tuneable dye laser as excitation source, with a Leitz MPV Compact microfluorometer. Pulses were about 100 ps long, enabling study of time-resolved fluorometry (see Chapter 12). A diffraction-limited excitation system enabled the measuring spot to be made as small as 0·3 μm. Emission was measured with a fast-response photomultiplier (rise time about 0·5 ns) and fed to an electronic signal averager.

van Geel *et al.* (1984) also described a microfluorometer with a pulsed nitrogen tuneable dye laser. Excitation was possible at 337 nm or in the range 357–710 nm by using different dyes in the laser. The spectral bandwidth was 0·1–3 nm. An area of the specimen 1 μm diameter was illuminated with short pulses of high intensity and the fluorescence measured with a fast detection system.

Spring & Smith (1987) described a system using an acousto-optic modulator to vary the excitation wavelength between several lines obtainable with an argon laser, and to obtain constant output intensity at all wavelengths.

Related instruments were described by Shoemaker & Cummins (1976), Cummins, Rahn & Rahn (1982) and Quaglia *et al.* (1982), who described microspectrofluorometers (see Chapter 7), and Kirsch, Voigtman & Wineforder (1985).

INDIVIDUAL COMPONENTS

Light sources

Arc lamps

Arc lamps are described in *Fluorescence microscopy* (Rost, 1991a). It remains to say only that, for quantitative purposes, stability is paramount. The subject was reviewed by Oldham, Patonay & Warner (1985). A magnetic device to stabilize the arc was suggested by Demko & Todd (1976) and one was at one time sold by Farrand (Farrand Optical Co. Inc., New York, USA). Oldham *et al.* (1985) reported that arc wander was reduced by adding an AC component to the DC arc supply. See also Chapter 4.

Lasers

A laser is essentially a device for generating very pure monochromatic light, ideal for fluorescence excitation. The light may be generated either continuously or in short pulses. For microfluorometry, brief pulses have the advantage of enabling measurements to be taken before there is time for photodecomposition to occur, and if the pulse is short enough (or at least ends sharply) the decay time of fluorescence can be determined (see Chapter 12).

A laser is an optical oscillator which emits light. Like other oscillators, it consists of an amplifier, a positive feedback mechanism to render the amplifier unstable and make it oscillate, a resonator to determine the frequency of oscillation, and a power source. A laser has three main components (Fig. 2.6): the *active medium*, the *pumping source*, and the *resonator*.

The active medium is a solid, liquid, or gas, which acts as an amplifier. It is a fluorescent substance such as a ruby crystal or a solution of a fluorescent dye. Amplification takes place by the phenomenon of stimulated emission. If a quantum of energy with frequency identical with that which would be emitted spontaneously impinges upon an excited atom, both quanta are emitted simultaneously and in phase with each other. Each of the two quanta is then capable of stimulating emission by other excited atoms, producing four quanta in phase, and so on. For amplification to occur, the medium has to be kept in a state of population inversion, in which the number of atoms in an excited state is greater than the number in a lower energy level.

The pumping source excites the active medium and is powerful enough to achieve this population inversion; power for the laser is provided by the pump. Solid-state (e.g. ruby) and liquid lasers are excited by light sources such as flash lamps or another laser; gas lasers are pumped by electrical discharges.

The optical resonator consists of a pair of mirrors facing each other and enclosing a space. When the active medium is placed inside the resonator, the system acts as an oscillator. The mirrors of the resonator provide positive feedback to the amplifier, to produce oscillation. One mirror is made semi-transparent to provide an output path for the laser light. The result is the output of a beam of light, which is characterized by high intensity, monochromaticity, and coherence. In some designs, the laser can be tuned over a range of wavelengths by changing the resonant wavelength of the resonator. This can be achieved by replacing one mirror with a diffraction grating; turning the grating alters the effective frequency of the grating and hence the resonant wavelength.

The light emitted by other light sources described in this book (arc and incandescent lamps) is radiated in all directions and over a wide range of wavelengths. Light from a laser, in contrast, is highly directional and monochromatic. The monochromaticity of the laser emission is determined by the nature of the active medium and the resonator. Under practical operating conditions the monochromaticity is limited by factors such as fluctuations in temperature and mechanical oscillations of the resonant cavity. If the emitted beam is too narrow for practical application, an increase in the width of the beam is achieved by using an optical system similar to an astronomical telescope in reverse. The same nomenclature for telescopes is applied – Galilean, Newtonian, etc.

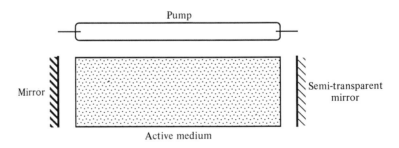

Fig. 2.6. Diagram of a laser, showing the active medium, pumping source (here a gas-discharge tube), and resonator (here two plane parallel mirrors facing each other; one is partially transparent to pass the output beam).

The ruby laser was the first successful laser design. Ruby consists of crystalline aluminium oxide (Al_2O_3) with a small amount of trivalent chromium (Cr^{3+}); it absorbs violet and green light at wavelengths around 400 and 550 nm, and fluoresces at 694·3 nm. A ruby laser contains a single crystal of ruby whose ends are flat, parallel, and completely silvered at one end and partially silvered at the other to form the resonator. The crystal is usually surrounded by a coiled flash lamp, which acts as the pump. Helium-neon, carbon dioxide and other gas lasers consist of a gas-discharge tube (like a neon sign) with a mirror at each end. Dye lasers employ solutions of fluorescent dyes in various solvents such as water, ethanol, methanol, or ethylene glycol. Many dyes have broad absorption and emission spectra, and are therefore suitable for tuneable lasers. Dye solutions have the additional advantages that they are cheaper than crystals, and that a wide range of fluorescent dyes is available.

Lasers have now been used in fluorescence microscopy for some years; the earliest references to such use that I have found are the papers of Kaufman et al. (1971), who described a scanning system, and Bergquist (1973), who used a laser for observation and photography of FITC-labelled antibodies. Lasers are now used mainly for quantification and for scanning techniques (see Chapter 15).

For further information on lasers, see: Thyagarajan & Ghatak (1981), Andrews (1986), Fleming & Beddard (1978), Herrmann & Wilhelmi (1987), Kaufman et al. (1971), Letokhov (1986), Schäfer (1977) and Steppel (1982). Laser dyes have been described by Maeda (1984), Drexhage (1973) and Rost (1991b).

Filters

Excitation filters must have a very strong cutoff at long wavelengths, and be stable under irradiation by the light source. Of glass filters, the standard combination of UGI and BG38 is quite suitable for selecting the 365 nm mercury line; otherwise a combination of glass and interference filters is best. For full stability under irradiation, a prism monochromator (Chapter 7) is much preferable.

Barrier filters can be either bandpass or barrier-type. Normal barrier filters, in my experience, are usually satisfactory; they give highest transmission of the emission spectrum, and are supplied as standard equipment to fit the microscope. Interference bandpass filters can be used if necessary to give increased optical specificity, at the expense of sensitivity.

It is essential that the optical system, including the barrier filter and the dichromatic mirror be non-fluorescent. Unfortunately some yellow and orange glasses used for barrier filters are fluorescent. If it is necessary to use a barrier filter which is fluorescent, the fluorescence can be substantially reduced by adding a second barrier filter with a shorter wavelength cutoff in front of the offending filter; this second filter will absorb a part of the excitation beam without significantly affecting the cutoff wavelength of the combination. Dichromatic mirrors showing significant fluorescence (Kaufman et al., 1971) must be replaced with non-fluorescent mirrors (for possible alternatives, see Chapter 7).

Objectives

The prime purposes of the objective are to provide resolution of the object and initial magnification. The brightness of the final image is also determined in part by the objective, according to the square of the numerical aperture (NA) and inversely as the square of the magnification. In epi-illumination, the objective acts as a condenser; and again its NA and magnification affect the brilliance of the illumination and therefore of the image, although in this case the brilliance is directly proportional to the square of its magnification, since this produces a corresponding demagnification of the light source.

Objectives are available in great variety, differing in magnification, NA, chromatic correction, flat-field correction, other corrections, the type of immersion fluid (if any), and other characteristics. Some are made specifically for fluorescence microscopy; many that are not can still be used very successfully. For quantitative work, minimal autofluorescence in the objective is essential.

The NA of the objective affects or determines the resolution of the system, the depth of field, and the brightness of the image. The effects are different in conventional and confocal scanning systems (see Chapter 15).

The resolution of the system depends primarily on the NA of the objective. When maximal resolution is required, the practical considerations involved are essentially the same as for ordinary microscopy, except that the NA of the condenser is no longer relevant. The fluorescence microscopist will in any case be using an objective with the highest convenient or available NA, to maximize the brightness of the image. Light losses may be greater in the more complex objectives, but this may be more than compensated for by increase in NA. A high NA also gives a narrow depth of field, which may or may not be advantageous.

Working distance of objectives is generally related to the NA, being least for objectives with highest NA. Some types of objectives are designed with a particularly long working distance (e.g. the new Olympus LB objectives), and are useful for dealing with thick specimens. Usually, difficulties with working distance arise only with immersion objectives.

The optical corrections of objectives have the same implications in fluorescence microscopy as in ordinary microscopy. Otherwise, in principle the more highly corrected objectives (apochromats) have a higher NA than achromatic objectives and therefore provide a brighter image, particularly with epi-illumination. However, the simplest achromatic objectives usually have fewer air–glass interfaces and no fluorescent glass, and may thereby achieve a higher light transmission and darker background. In my experience, flat-field objectives seem to be particularly suitable for epi-illumination, probably because they tend to have an large rear pupil which accepts the full beam of light from the epi-illuminator.

Ultraviolet-transmitting objectives are only required for epi-illumination with particularly short wavelengths (below 360 nm). They are essential for microfluorometry of some fluorochromes (see Chapter 5) and for determination of UV excitation spectra (see Chapter 6), but are not required for routine purposes; just as well, because they are very expensive!

Objectives for uncovered specimens, corrected for use without a coverslip, are intended for use with uncovered smear preparations.

Immersion objectives are of two types: those intended for use with conventional coverslipped slides, and low-power water-immersion objectives. The former usually exploit the possibilities of increased NA as a result of increased refractive index of the medium between the objective and specimen. The immersion medium may be conventional non-fluorescent immersion oil, silicone oil, glycerol, or water. Water-immersion objectives can be used for studying objects under water in a vessel without a lid.

The diameter of the objective's exit pupil is of importance in epi-illumination, because this is also the entrance pupil for illumination, and must be filled by an image of the light source. If this image is smaller than the pupil, uneven illumination will result; if the image of the light source is greater than the pupil, light is wasted and the intensity of the fluorescence correspondingly reduced. The diameter of typical objectives varies from about 12 mm (low powers) to about 5 mm (high powers).

Field diaphragms

At least two field diaphragms are required. The illuminating field diaphragm corresponds to the normal field diaphragm of any fluorescence microscope. If a reference channel is employed (see Chapters 7 and 9), there may need to be a corresponding diaphragm in the reference channel. The measuring field is a region of the field delineated by a field stop (diaphragm) in the magnifying system, within which measurement is made.

Diaphragms used for limiting the fields of illumination and measurement are of three types: iris, rectangular and fixed. The best arrangement is a set of interchangeable diaphragms allowing all possibilities. Iris diaphragms are circular, and have but a single adjustment (for diameter), so they are simple to operate. However, a circular field is not always appropriate. A rectangular adjustable diaphragm has two slit-like mechanisms at right angles, so that length and width can be adjusted separately. Fixed diaphragms consist of a metal plate, incorporating a hole of suitable dimensions. These are most conveniently made circular, but in principle may be of any shape. The main advantage of this type is consistency: the size and shape remain constant. It is convenient if all diaphragms in the apparatus are standardized to a single fitting, so that all diaphragms are interchangeable. It is also necessary that diaphragms can be centred to the optical axis of the instrument.

Temperature-controlled stages

We (Rost & de Permentier, unpublished) have successfully used the Leitz heating and cooling stage on a Leitz Orthoplan microscope in conjunction with the Leitz Kryomat refrigeration unit to maintain temperatures in the range 10–40 °C. Tiffe (1977) has described a cooling system for cytofluorometry in the range 3·5–300 K.

Photodetectors

The measurement of light requires a light-sensitive device which produces an electrical signal related to the intensity of the light. At the present time, such light-sensitive devices fall into two classes: vacuum devices (such as photomultiplier tubes and television cameras), and solid-state devices (photoresistors, phototransistors and charge-coupled devices).

Photoelectric cells, which were the forerunner of photomultipliers, contain, in a vacuum, a photocathode and a positively charged anode; when light shines on the photocathode, a current will pass between the electrodes and can be measured. The strength of the current is proportional to the intensity of the light. The relationship between the wavelength (quantum energy) of the light and the intensity of the current may be complex. In principle, at any given wavelength, a certain proportion of incident photons may be expected to be successful in releasing an electron. The proportion of successful to total photons is known as the quantum efficiency of the photoelectric cell. This efficiency will vary with the wavelength. Below a certain level of quantum energy, the efficiency will be zero and no current will flow; photomultipliers are therefore insensitive to red. At the other end of the spectrum, the efficiency of the photoelectric cell will be limited by the transparency of the envelope: ordinary glass will not transmit UV, while quartz will. The efficiency of the device, particularly at long wavelengths, can be increased by selecting a suitable material for the cathode. In practice, a metallic cathode is coated with a compound of one or more rare earths.

Photomultipliers

A photomultiplier tube (PMT) is an electronic device for measuring weak light. Its output is in principle a series of pulses, each corresponding to an incident photon; if there are sufficient pulses per second these add up to an appreciable current. The device requires a high-voltage DC power supply, at about 600–1200 V, which must be very highly stabilized.

Photomultipliers have a metallic cathode coated with a compound of one or more rare earths. Incident photons knock out electrons from the cathode, which are attracted towards the first of a series of positively charged elements called dynodes. As a photon releases an electron from the cathode, the electron is accelerated towards the first dynode because of the dynode's positive potential (typically about 100 V with respect to the cathode). The electron, on hitting the first dynode, will have attained a higher energy level than it had on leaving the cathode, because of having been accelerated through a potential of 100 V or so. The electron will therefore have enough energy, on hitting the dynode, to knock out several electrons. These are now attracted to a second dynode, held at a higher positive potential (another 70–100 V; now about 200 V with respect to the cathode). In consequence, an amplified current will flow towards the second dynode. The process is repeated at several dynodes, so that, although only a single electron may be released from the cathode by one photon, the result is a substantial pulse of current at the final anode. The increase in sensitivity is enormous. The amplification of each dynode depends on the coefficient of secondary emission (which generally lies between 3 and 7) and the efficiency of transfer from one dynode to the next. The average value of amplification of current per stage is about 4·5. Hence, a typical photomultiplier with 9 dynodes has a gain of about $(4·5)^9$, i.e. approximately one million.

With no light incident on the photomultiplier, there is still a small amount of emission from the cathode, mainly thermionic (i.e. associated with the temperature of the cathode). This results in a small output, the so-called dark current. The amount of the dark current varies to some extent from one photomultiplier to the next, and is much reduced by cooling and by keeping the photomultiplier in a dry environment.

Photomultipliers have a glass envelope with a flat window in that part of the glass tube which is over the cathode. This window may be either at the end of the tube or at the side near one end, giving rise to the 'end-window' and 'side-window' types, which require different shapes of housing. The window is normally of glass, but can be made of UV-transparent material (quartz-glass) if required. There is of course no point in paying money for a quartz window if only visible light is to be measured. Photomultipliers are sold in different grades (A, B) according to quality, and can be specially selected for characteristics such as high sensitivity or low noise. For further details of photomultipliers see O'Connor & Phillips (1984), Candy (1985a,b) and manufacturers' data (e.g. EMI catalogue).

The output from the photomultiplier can be measured in either of three basic ways: DC current measurement, AC current measurement in conjunction with a light chopper, and photon counting.

DC current measurement measures the aggregate of the pulses due to the individual photons. If the pulse rate is fast (say in excess of 100 000 pulses/s) it is easiest to measure the total current. This current is of the order of microampères, and can be measured in any of the ordinary ways appropriate to a current of that level in a high-impedance circuit. Appropriate instrumentation includes high-impedance chart recorders (e.g. see Gillis *et al.*, 1966) and digital voltmeters, usually placed in parallel with a load resistor of a megohm or so. An analogue-to-digital converter can be used: this initially converts the measured current to a series of pulses of constant voltage whose frequency varies as the strength of the input, which are then counted (e.g. Freitas, Giordano & Bottiroli, 1981). DC current measurement in general is the simplest method, but its accuracy is critically dependent on the stability of the high-voltage supply for the photomultiplier.

A mechanical chopper in the excitation light path reduces the effect of stray light in the system. The photomultiplier signal is amplified by an AC amplifier with a filter to favour the frequency of the chopper. Constant stray light, not modulated by the chopper, is therefore ignored by the system. Also, an AC amplifier tends to be easier to construct and cheaper than a DC amplifier.

Photon counting is the best method (in my opinion) for measurement of very low light levels, and finds its major application in astronomy. In this method, the pulses associated with each photon are amplified, passed through a discriminator which eliminates the weakest pulses (which are mainly noise from the photomultiplier tube), and then counted during preset intervals. Besides giving a direct digital output, photon counting has relative immunity to variations in the high-voltage supply to the photomultiplier (which affects the height of the pulses but not their number), the temperature of the photomultiplier, and the position of the light spot on the photocathode. On the other hand, pulse-counting circuits are susceptible to radio-frequency interference, and corrections have to be made for pulse pile-up; the electronics cannot distinguish pulses which arrive too close together. If a counting device which is counting random pulses has a fixed finite resolving time, some pulses will be lost because they arrive while the circuitry is still working on the previous pulse. If the counter is insensitive during a time p, if the observed count is N per second the best estimate of the true mean pulse frequency is $N/(1 - Np)$. This is discussed in more detail in Chapter 4. At very high count rates, the corrections become too great to be satisfactory. Present-day instrumentation can count up to about 10^6 pulses per second, with a dead-time (period after each pulse during which the circuit cannot accept another pulse) of about 60 ns (Hollis, 1987). Photon-counting circuits for microfluorometry were described by Pearse & Rost (1969), Rost & Pearse (1971), Cova *et al.* (1974), Wittig, Rohrer & Zetzsch (1984) and Candy (1985a,b); see also Kemplay (1962).

Video cameras

These are generally more suitable for scanning than for simple measurements from a specified field, but offer interesting possibilities for measurements from irregularly shaped fields using image-analysis equipment. The technique is known as quantitative video intensification microscopy (QVIM), and is described in Chapter 15.

Solid-state devices

It is only very recently that solid-state photo-detectors have become able to compete with photomultiplier tubes as regards sensitivity. A photodiode (photoresistor) is a device with a high resistance in the dark, and a lower resistance when illuminated. Stoward (1968b) used a photodiode, for measurement of rather strong fluorescence. I have used such a device in a reference channel to monitor the brightness of the excitation. Probably, in the future, charge-coupled devices (Hobson, 1978) will be used.

AUTOMATION

It is possible to arrange a series of electrically operated shutters to control the light paths for the transmitted-light, epi-illumination and photometric pathways, and to arrange that a single command operates them in sequence and triggers the measurement and its logging. For multi-wavelength studies, the excitation filter, dichromatic mirror and barrier filter appropriate to each wavelength can be selected by the controlling microprocessor. Automation has several advantages: commonly, the most important benefit is saving of time during the measurement process, minimizing fading due to irradiation. Also, errors are less likely to be made in an automated system. Automatic data logging saves much time and reduces the risk of human error in transcribing data. The Leitz MPV MT2, Olympus BH2-QRFL and Nikon P1 systems (see above) are or can be extensively automated, as is the Joyce-Loebl MagiCal system, described in Chapter 15. Pneumatic rather than electric components were employed by Haaijman & Wijnants (1975). Other reports of automation include those of Combs (1973), de Josselin de Jong, Jongkind & Ywema (1980), Docchio *et al.* (1984), Enerbäck & Johansson (1973), Gordon & Parker (1981), Neely, Townend & Combs (1984), Reuter (1980), Rost & Pearse (1971), Ruch & Trapp (1972), Rundquist & Enerbäck (1985), Seul & McConnell (1985), and Wasmund & Nickel (1973).

3

Microfluorometric technique

This chapter reviews the practical aspects of making a quantitative measurement by microfluorometry. It is assumed that the object of the exercise is to determine the amount of a fluorescent substance present in a specified area of the specimen. The process of measuring fluorescence emission consists of four stages: setting up the instrument and selecting appropriate excitation and emission wavelengths, locating the specimen and defining the area from which measurement is to be made, determining the fluorescence intensity of both the test specimen and a standard, and calculating the result. In designing the microfluorometric procedure, it is necessary to take into consideration the necessity to avoid sources of error such as fading, and sometimes special considerations such as a need to measure at a controlled temperature. Before making any microfluorometric measurements, it is well to spend some time and thought on designing the experiment.

Descriptions of other workers' techniques are given in the papers referred to in Chapter 5; particularly good descriptions of procedure are given by Alho, Partanen & Hervonen (1983) and by Fabiato (1982).

Setting up the instrument

Before any measurement can be made, the microfluorometer must be set up properly, its performance checked, and any necessary adjustments made to minimize errors (see Chapter 4). The setting up of a microfluorometer involves adjusting the instrument as with any other fluorescence microscope, except that greater care must be taken, and additional photometric equipment must be operational.

The setting up of the illumination system and of the microscope is described in *Fluorescence microscopy* (Rost, 1991a). For quantification, this process is even more important than it is for qualitative examination, and every care must be taken. An arc lamp to be used for excitation (xenon or mercury lamp) needs to be switched on 30 min before a measurement is made, in order to enable it to attain stable operation. The microscope should be checked item by item, commencing at the light source(s) and finishing at the detector. Points to watch include cleanliness of the optics (essential), centration of the light source and illumination system, and the focus of the collector lens in the lamphouse. If a filter is used for monochromation, ordinary Köhler illumination will be used. If a monochromator is used in the excitation system, special care is required in setting up the illumination (see Chapters 7–9).

Checking

The validity of the measuring system must be checked by tests for stability, uniformity of illumination and linearity of measurement.

Stability of light source

The procedure for checking the stability of the light source is normally as follows. First, the lamp must be allowed a sufficient warming-up period; about 30 min should be enough. A non-fading object is placed on the stage of the microfluorometer, and the microscope focussed on it (see Chapter 4, Standards for fluorometry). A thick piece of uranyl glass is most commonly used. Possibly better is a solution of a fluorochrome; assuming epi-illumination, there should be no difficulty in placing a small vessel containing a fluorochrome solution under the objective. The advantages of a solution are a choice of fluorophore and a low rate of fading. Alternatively, a mirror can be used, the measurement being made at the 'excitation' wavelength. (Caution: the light may be extremely bright!) The output of the measuring device, with reference channel disconnected, is connected to a chart recorder and the system adjusted until a suitable deflection on the recorder is obtained, say half to two-thirds full scale. The recorder should be set to run at a fairly low speed, e.g. 15 mm/min. This method presupposes stability in the measuring system; this in turn can be checked by a standard lamp supplied from a stabilized DC source. There should be no significant variation in the output. Alternatively, in instruments provided with a reference channel, the chart recorder can be connected to the output from the reference channel. This avoids the need for a non-fading object, or indeed any object at all.

Homogeneity of illumination

It is essential to check for homogeneity of illumination of the measuring field in terms of both intensity and wavelength; a similar criterion applies in respect of the measuring field. In a microspectrofluorometer (see Chapter 7) the emission spectral measuring system can be used to check on the spectral distribution of the excitation light. The attainment of homogeneous illumination in the object plane depends mainly on (1) correct alignment of all optical components to the optical axis of the illumination system, (2) correct focussing of the collector lens (this varies with wavelength, unless the collector lens is highly corrected), and (3) correct focussing of the projection lens to image the exit slit of the monochromator into the aperture of the condenser (i.e. the 'exit pupil' of the objective, when epi-illumination is used).

Centration of the lamp and adjustment of the collector lens are checked by examining the pattern of a white card held in the light path after the projection lens. The projection lens is adjusted by focussing an image of the monochromator exit slit onto a white card placed approximately 5 mm above the level of the object plane, with no objective in place: this plane is approximately that of the exit pupil of a high-power objective.

Rigler (1966), who first introduced a test for homogeneity of field, employed a small glass 'crystal' (a very small fragment) of uranyl glass, or a lymphocyte stained with Acridine Orange (the fluorochrome with which he was working). The latter is less suitable than glass because of fading, although the use of the same fluorophore as one is measuring is generally to be preferred because homogeneity is being checked under exactly the same conditions as will be used for the measurement, including the excitation and emission spectra of the fluorophore. Böhm & Sprenger (1968) in quantifying DNA by the fluorescent Feulgen method, used stained bull sperm nuclei, measuring the fluorescence of a nucleus at each of five positions in the field (centre, top, bottom, left and right). A similar 'test of the five positions' had been employed in absorbance microspectrophotometry by Garcia (1962) and Garcia & Iorio (1966). Böhm & Sprenger had to use a different nucleus at each position, because of fading. Their published results showed a coefficient of variation (s.d. × 100/mean) of 2·2.

Another method is the 'nine-position' test (Fig. 3.1) of Böhm & Sprenger (1970). In this test, the object is uniform, such as a sheet of uranyl glass, while the iris diaphragm of the measuring field is moved in relation to the illuminated field. They used a small opening in a diaphragm placed in front of the photomultiplier, corresponding in diameter to 6·4 μm in the object plane. The eyepiece was provided with graduated cross-hairs in X- and Y-axes; the measuring diaphragm was moved successively to each of nine positions in the measuring field. They were able to obtain a coefficient of variation of 2·0.

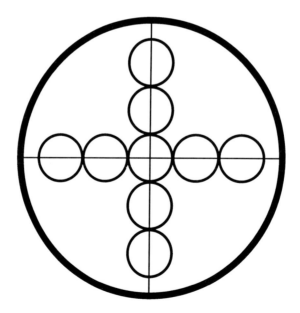

Fig. 3.1. The 'nine-position' test of Böhm & Sprenger (1970). The nine different positions, along cross-hairs, for assessment of homogeneity of illumination and measuring field (after Böhm & Sprenger, 1970).

I prefer a test in which a stable fluorescent particle, of uranyl glass or plastic scintillator, is measured in a similar set of nine (or more) positions, the specimen being moved while everything else is kept constant. This ensures that the test is made under exactly the same conditions as a normal measurement. An interesting example is shown in Fig. 3.2, made with the microfluorometer of Rost & Pearse (1971; see Chapter 16) using a monochromator in the illumination system; the figure shows the difficulty in obtaining uniform illumination at right angles to the slit.

Homogeneity of measuring field
Exactly the same considerations as for even illumination of the field apply to the photometric detector's view of the measuring field. This means that a spot of given luminance should have the same effect on the measuring device irrespective of its position in the field. This also may be tested using the 'nine-position' test of Böhm & Sprenger (1970) described above. In fact, the fluorescent particle test described in the previous section will simultaneously test for inhomogeneity of both illumination and measurement systems.

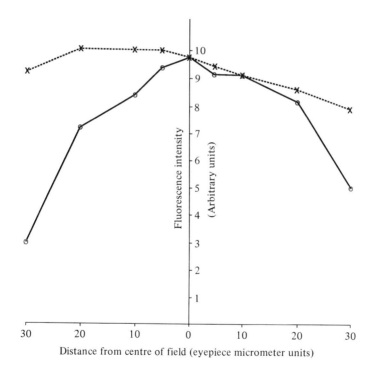

Fig. 3.2. Measured brightness of a stable fluorescent particle (BBL test slide) at various positions along diameters of a measuring field in a microspectrofluorometer having a monochromator in the illumination system. Using epi-illumination, the exit slit of the monochromator was focussed by a projection lens into the 'exit' pupil of the objective. Crosses: measurements along the diameter parallel to the mono-chromator slit. Circles: measurements along the diameter perpendicular to the slit.

Light leaks

To search for light leaks into the optical system, darken the room and shine light from a torch successively on all possible places, starting at the photomultiplier housing. Look particularly for leaks around the control levers of the iris diaphragm in front of the photomultipliers. These can be dealt with by keeping black cloth draped over this part of the apparatus.

In my experience, it appears to be necessary not only to check the apparatus carefully for leaks but also to operate in a darkened room with a black cloth over any vulnerable parts or with a cabinet built around them, and with no direct illumination of the microfluorometer. A black cloth should be draped over the microscope during measurement. Stray light can also be reduced by closing the aperture diaphragm of the substage condenser when this is not in use for illumination; better, the condenser can be replaced with a light-trap, which is basically a cup-like container lined with matt black material and preferably shaped so that no surface can reflect light back up into the microscope. If this is not available, the condenser can be lowered and a piece of non-fluorescent black cloth or cardboard placed over it. All eyepieces must be covered or blocked off.

Linearity

A check for linearity of measurement is most easily done by use of a non-fading test object and a series of neutral density filters of known transmission (Zeiss supply excellent ones). Linearity in the reference channel is checked by placing filters (the exact density need not be known) in the illumination pathway to reduce the brightness of the illumination; the final calculated reading for a stable test object should not change.

Selection of wavelengths

Other things being equal, the excitation and measuring wavelengths should correspond as nearly as practicably to the excitation and emission maxima of the fluorophore to be measured. In practice, the filter combination suitable for visualization of the fluorophore by fluorescence microscopy is very likely to be adequate for quantification too. However, filters with narrower bandwidth will give increased optical specificity, with only some loss of sensitivity due to reduced transmission.

On the excitation side, the illumination system is set up for illumination at the desired wavelength, which (for highest specificity) should be at or near the excitation peak of the fluorophore. Sometimes a narrow-band excitation from a laser or mercury arc line a little removed from the optimal wavelength is to be preferred to a broader-band excitation at the exact peak wavelength, as might for example be obtained from a xenon arc lamp and an interference filter. Also, sometimes it may be necessary to increase the separation of the excitation and measurement wavelengths in order to reduce stray light. Sometimes also, interference from autofluorescence or other unwanted fluorophore can be reduced by a judicious shift of either the excitation (preferably) or measuring wavelength, or both. If a monochromator is used, the wavelength is set, the monochromator slits adjusted to the desired width and an appropriate supplementary filter chosen. The sensitivity of the reference channel (if any) is suitably adjusted to cope with the anticipated range of intensities during measurements; this is best checked by doing a dummy run with the light path to the specimen blocked to prevent fading.

The next problem is to decide the wavelength band in which to measure the emission, and how best to select that band. Selection of the wavelength range for emission measurement may be carried out in a number of different ways. In principle the emission may be monochromated by either a narrow-band device such as a monochromator or an interference filter, or by a broad-band device such as a barrier filter. A narrow-band device can be centred either on the emission peak of the substance to be determined if known or on a wavelength somewhere in the long-wavelength tail of the emission spectrum. If a barrier filter is used it may be designed to pass either the major part of the emission or only a region of the long-wavelength tail. Combined together these possibilities give a total of four basic systems.

On the emission side, the choice usually lies between a conventional barrier filter or a bandpass interference filter. A narrow-band emission filter centred on or near the peak emission wavelength of the fluorophore gives the greatest possible degree of specificity. It also gives greater sensitivity than a filter of equal transmission and bandwidth measuring in the long-wavelength tail where the intensity of emission is less. On the other hand the transmission curve of the filter may overlap with the excitation filter, resulting in a high background reading.

A filter which has a narrow bandwidth and peak transmission somewhere in the long-wavelength tail of the emission is the least sensitive of the four systems; such a filter has the double disadvantage of a narrow bandwidth and measurement in a wavelength region where the emission is less intense than at the peak. On the other hand, this system gives a somewhat higher degree of specificity than the use of a barrier filter with the same cut-off on the short-wavelength side. If an interference filter or a monochromator is used for this purpose it may suitably be combined with the barrier filter to give a sharper cut-off and therefore less overlap with the excitation.

A barrier filter transmitting almost all of the emission gives the greatest possible sensitivity, since the transmission of the filter is high and nearly all the emission spectrum is transmitted. However, even if the barrier filter itself has good rejection characteristics at short wavelengths there will be still be a high likelihood of overlap of its transmission spectrum with that of the excitation filter. Of course, if a laser is used for excitation the latter problem does not arise. A barrier filter with a cut-off wavelength somewhat longer than that of the emission peak may provide a good compromise between sensitivity and non-overlap with the excitation filter.

Locating the specimen

To avoid fading, the desired region of the specimen should, if possible, be located without excitation of fluorescence, e.g. by conventional transmitted light microscopy, or by phase-contrast, interference-contrast, or darkground illumination. Use of red light minimizes photodecomposition. On the other hand, if emission is being measured in the red (e.g. from Pararosaniline), weak green light may be convenient for setting up and need not be switched off, since it will be blocked by the barrier filter (Giangaspero *et al.*, 1987). If the required area can be found only by fluorescence, the light used for excitation should be reduced in intensity by means of a neutral-density filter. Under these latter circumstances it may be helpful to have two sources for fluorescence excitation: one giving weak illumination over the entire field, as in an ordinary fluorescence microscope; and the other for measurement only, with a limited field. In any case, the illumination field diaphragm should be limited in size as much as possible to avoid fading more of the specimen than necessary. When the desired region of the specimen has been found, an image of the exact area to be measured must be placed within the measuring diaphragm. Before measurement commences, any light used for locating the specimen must be shut off.

Measurement

To commence measurement, a shutter in the excitation pathway should be opened and the emission measured. I have found it helpful to include a camera-type shutter in the excitation pathway: the flash synchronization contact (X-type, as for electronic flash) can be used to initiate measurement. Ideally, the whole procedure, including the shutter, should be under automatic computer control.

Background

There is invariably some background light, not forming part of the emission spectrum of the fluorophore, due to tissue autofluorescence and stray light in the system. The background (blank) should therefore always be measured. The blank must of course be measured under the same optical conditions as for the test; in particular, without changing the measuring diaphragm. Ideally, a non-fluorescent piece of tissue similar to the specimen should be used for this purpose; my experience is that this often actually gives a lower reading than a blank field, since it may block light reflected from below. Background can be reduced greatly by replacing the substage condenser (assuming epi-illumination) with a light-trap (see above).

Standardization

Standardization (calibration) involves the measurement of a fluorescent standard (see Chapter 4) under the same optical conditions as the test specimen. In the course of a session of microfluorometry, two kinds of standard may be measured: absolute standards of the fluorophore to be quantified, and a working standard. The former would be measured at least once, preferably twice (at the beginning and end of the session); the latter, which might be represented by a piece of uranyl glass, would be measured frequently as a check on the stability of the system.

Calculating the result

The final result (x) is calculated by the formula:

$$x = \frac{\text{(nominal value of standard)} \times [\text{(reading of test)} - \text{(reading of test blank)}]}{\text{(reading of standard)} - \text{(reading of standard blank)}}$$

If a reference channel is used, each of the readings in the above formula must be divided by the simultaneous reading of the reference channel (less the value of the reference channel's blank, which is ordinarily a negligible dark current). Data processing and statistical analysis are discussed in Chapter 4.

General considerations

Sensitivity

Fluorometry by its very nature tends to be more sensitive than microdensitometry, but (if only for this reason) tends to be used for measurements of very small amounts of material. In general, an increase in overall sensitivity can be achieved in two main ways: increasing the brightness of the illumination, and increasing the sensitivity of the detector. The second is far preferable, as any increase in illumination intensity leads to more rapid fading. Also, a bright light source gives rise to problems with heating, damage to excitation filters, and ozone generation. If at first the sensitivity of the instrument is inadequate, the following should be considered. An increase in numerical aperture of the objective will increase the efficiency with which fluorescence is collected; the intensity of illumination (assuming epi-illumination) will be increased but this can if desired be decreased again by placing a filter in the excitation pathway.

Measurement of red fluorescence (e.g. from Propidium iodide) may present a problem due to the low red-sensitivity of ordinary photomultipliers (see Chapter 2). A red-sensitive photomultiplier may be necessary.

Avoidance of fading

Fading during irradiation is inevitable. There are four ways of dealing with the problem: (1) measure with a low excitation level, and high detector sensitivity; (2) measure very quickly, before the object has had time to fade; (3) measure after a standard interval; and (4) make several measurements at known times after commencement of irradiation, plot these, and extrapolate backwards to an estimated intensity at time zero. Each of these methods has advantages and disadvantages.

A method in which a measurement is made almost instantaneously is generally to be preferred. A very short laser pulse can provide sufficient excitation for measurement without appreciable fading. Measuring quickly, before there has been much fading, requires specially designed equipment. Generally, the object must be located by relatively dim light, preferably red, possibly using phase-contrast or interference-contrast optics. A shutter is opened to allow the excitation to pass to the object (or a laser pulse is activated), and simultaneously the measuring system is triggered and a reading taken as quickly as possible.

Measuring after a standard interval is based on two premises: first, that the rate of fading is greatest initially, and that after a short time the rate of fading is much reduced; and, second, that the amount of fading is constant during the pre-irradiation before the measurement. This latter requirement may be difficult or impossible to justify. The fading rate will depend, in part, on the chemical characteristics of the mounting medium, such as pH, whether aqueous or non-aqueous, and also on the temperature.

Making several measurements at known times after commencement of irradiation, and extrapolating backwards to calculate the supposed intensity of the fluorescence at the moment of commencement of irradiation (Fig. 3.3), is a method which I introduced. It can be implemented by a chart recorder, or by digital readouts. In the latter case, ideally one should have an automatic data logging system, controlled by a clock, to make and record readings at preset intervals. I found that the main difficulty of using this method was that the rate of fading was never exponential, and it was always

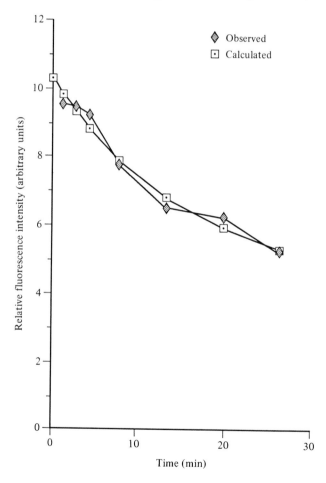

Fig. 3.3. Measurements of fluorescence, showing fading (diamonds). An exponential decay curve has been fitted (squares) and extrapolated back to the commencement of irradiation (time zero).

difficult to extrapolate backwards with any confidence. In principle, attempts can be made to fit more complex curves, e.g. the sum of an exponential function and of a constant term; or the sum of two exponential functions and a constant term. These usually (but not invariably) give a reasonable fit; in my experience the labour and computing power required is not justified by the results.

On the other hand, if a chart recorder is being used in any case to record the data (as is sometimes done) then it is little additional labour to extrapolate backwards by eye. The method in such case is that the specimen is located by phase-contrast or a similar method, the light for that is switched off, the chart recorder is switched on, the shutter to the photometer is opened, and finally the shutter to the excitation is opened. The pen of the recorder should immediately rise, and show a maximum within a second or so, followed by a decay characteristically fast at first and then slowing. By inspection, the trace is extrapolated backwards to the time at which the pen started to rise. The correction required should only be a small percentage of the peak reading; if not, the excitation is too bright, the excitation filter is inappropriate, or the fluorophore is fading too rapidly to be suitable. It can be argued that in this method, even if the method used for extrapolation is not perfect, any error in extrapolation is a second-order error and can safely be ignored.

Measurements at controlled temperature

It is sometimes necessary or useful to measure at a known or constant temperature, particularly for studying living tissues. I have used the Leitz heating-and-cooling stage, which fits on a normal Leitz rotating centring stage, replacing the rotating top. Heating is electric, cooling by a circulating liquid; temperature is controlled by a thermostat controlling the heating. The cooling liquid can be circulated from a tank of cold water, or can conveniently be supplied by the Leitz Kryomat unit. A difficulty with cooled stages is that water condenses on them, and may condense on the specimen. This is of little significance if a water-immersion objective is used. A thermistor can be incorporated into the specimen mount for accurate temperature measurement (see Chapter 13). Microfluorometry at low temperatures has been described with apparatus operating down to 77 K (Tiffe, 1975) and even 3·5 K (Tiffe, 1977). At these temperatures, phosphorescence would be likely to become important (see Chapter 12).

Care of the instrument

Care of the fluorescence microscope is described in *Fluorescence microscopy* (Rost, 1991a). It is essential to keep all optical components clean, to reduce light scatter and autofluorescence of dust particles. This is helped if the instrument as a whole is kept dust free with a vacuum cleaner or damp cloth; avoid sweeping, which puts dust into the air ready to settle on optical surfaces. A miniature battery-operated vacuum cleaner is available for cleaning small objects, optical surfaces, etc., although it must be admitted that it is not quite as effective as a blower which can produce a jet of air under higher pressure than a vacuum cleaner can (the latter is limited to a pressure of 1 atm).

The photomultiplier must be handled only with great care. It should not be subjected to bright light (especially not with the high voltage on, in which case it will be damaged permanently); a period of exposure to light increases the dark current in subsequent measurements. My own practice is to darken the room as much as is practicable before opening the photomultiplier housing, and to store any spare photomultiplier in a light-tight box. The envelope over the cathode is an optical structure, and it should be treated accordingly. The photomultiplier must be kept dry at all times; damp provides leakage paths for the anode current, reducing the output and introducing noise. Silica gel in the photomultiplier housing, away from the high voltage circuits, may dramatically improve performance (Miles, 1986). If silica gel is used, it must be checked or changed at regular intervals.

4

Microfluorometry: errors, standardization and data processing

Measurements by microfluorometry are subject to errors of three main types, related respectively to theoretical considerations, the specimen and to the apparatus. It is important to distinguish between *precision* and *accuracy*. A measurement may be very precise, e.g. to four decimal places, without being particularly accurate (say only to two decimal places). The precision of a measurement depends only on the readout device, whereas accuracy is determined by the experimental conditions as a whole. The specificity of the method also has to be considered.

Basic errors

Relationship between measured fluorescence intensity and amount of fluorophore

All microfluorometry is based on the concept of irradiating a specimen with nominally monochromatic light of known intensity, and measuring the intensity of light emitted from a given area of the specimen. The quantitative relationship between the intensity of the fluorescence and the amount of fluorescent substance present has been discussed by Böhm & Sprenger (1968), Fukuda, Böhm & Fujita (1978), Oostveldt *et al.* (1978), and Gains & Dawson (1979).

It is generally assumed that the intensity of fluorescence is proportional to the amount of fluorophore present. Unfortunately this is true only at low fluorophore concentrations, as will be shown below. The strength of the emitted light is dependent on two factors: the amount of light absorbed, and the proportion of the absorbed light re-emitted as fluorescence. It is convenient to consider quantitative aspects of each stage separately. In what follows, the fluorophore is assumed to be uniformly distributed in the region measured, scattering of light is neglected, and the objective is assumed to have a low numerical aperture as for microdensitometry.

The following symbols will be used:

Amount of fluorescent substance	x
Intensity of the incident light	I_0
Intensity of the absorbed light	I_a
Intensity of the transmitted light	I_t
Intensity of the emitted (fluorescence) light	I_f

41

The ratio, I_f/I_0 F

Concentration of the fluorophore c

Depth (thickness) of the specimen d

Area of the measured field a

Molar extinction coefficient of the fluorophore ϵ

In densitometry, the extinction is calculated from observed values of I_0 and I_t. In microfluorometry, the observable data are the fluorescence intensity of the specimen (I_f) and of a standard containing a known amount of the fluorophore. In effect, we require a means of calculating the extinction, and hence the fluorophore concentration, from these data.

The intensity of the measured fluorescence (I_f) is equal to the amount of light absorbed by the specimen (I_a) multiplied by the quantum efficiency of the fluorophore (ϕ) and further multiplied by a constant factor (r) dependent on the instrument, i.e.:

$$I_f = I_a \phi r \tag{1}$$

The amount of light absorbed (I_a) is equal to the difference between the amount of light incident on the specimen (I_0) and the amount of light transmitted (I_t). The amount of light transmitted is given by the Beer–Lambert law,

or:
$$\log (I_t/I_0) = -\epsilon cd$$
$$I_t = I_0 e^{-2\cdot 3 \epsilon cd} \tag{2}$$

where $2\cdot3$ is an approximation for the natural logarithm of 10 (more accurately $2\cdot302$). Combining the above equations, we obtain:

$$I_f = I_0 \phi r(1 - e^{-2\cdot 3 \epsilon cd})$$
$$I_f/I_0 = \phi r(1 - e^{-2\cdot 3 \epsilon cd})$$

It is convenient to denote I_f/I_0 by a symbol, F, representing the relative fluorescence intensity, i.e. the fluorescence intensity for a constant light input. We now have:

$$F = \phi r(1 - e^{-2\cdot 3 \epsilon cd}) \tag{3}$$

The amount (x) of fluorophore present is determined by the concentration (c), thickness (depth, d), and area (a) of the measuring field, according to the relation $x = acd$. Substituting x/a for cd in the previous equation, we have:

$$F = \phi r(1 - e^{-2\cdot 3 \epsilon x/a})$$

This, the true relation between x and F, is too complex for general use. However, for practical purposes, approximations can be made to reduce the equation to simpler forms. The exponential expression in equation (3) can be expanded to give:

$$F = \phi r \left[1 - \left\{ 1 - \frac{2\cdot 3 \, \epsilon cd}{1!} + \frac{(2\cdot 3 \, \epsilon cd)^2}{2!} - \frac{(2\cdot 3 \, \epsilon cd)^3}{3!} + \; \ldots \; + (-1)^n \frac{(2\cdot 3 \, \epsilon cd)^n}{n!} \ldots \right\} \right]$$

whence:

$$F = \phi r \left[\frac{2\cdot 3 \, \epsilon cd}{1} - \frac{(2\cdot 3 \, \epsilon cd)^2}{2} + \frac{(2\cdot 3 \, \epsilon cd)^3}{6} + \; \ldots \; + (-1)^{n+1} \frac{(2\cdot 3 \, \epsilon cd)^n}{n!} \ldots \right] \tag{4}$$

For low values of ϵcd, if the cubic and higher terms are ignored, the equation can be simplified to:

$$F = \phi r \left[2 \cdot 3 \; \epsilon cd \left(1 - \frac{2 \cdot 3 \; \epsilon cd}{2} \right) \right] \qquad (5)$$

This approximation is valid to within 4% provided that ϵcd is less than $0 \cdot 17$ (Gains & Dawson, 1979).

For very small values of ϵcd, second and higher powers can be neglected, and equation (4) reduces to the linear relation:

$$F = \phi r \, (2 \cdot 3 \; \epsilon cd) \qquad (6)$$

In other words, the emitted fluorescence becomes directly proportional to the concentration of the fluorophore. If the amount of the fluorophore present is other than very low, the relationship is non-linear. Table 4.1 shows the relation between the extinction and the absorption, and also the ratio of these two. At very low extinctions, the ratio absorption/extinction is about $2 \cdot 30$; as the extinction increases, so the ratio drops, which means that the amount of substance present is underestimated by microfluorometry (if a linear relationship between light absorbed and fluorescence is assumed). As a rough guide, the amount of the error is roughly equal to the extinction; so that, if the result is to be obtained within 1% accuracy, the extinction should not exceed $0 \cdot 01$, while for a 10% accuracy the extinction should not exceed $0 \cdot 1$. It has been suggested that, for practical purposes, the relationship can be taken as linear provided that the extinction of the fluorophore does not exceed $0 \cdot 2$ (Böhm & Sprenger, 1968).

The above theoretical discussion is based on the assumption that the numerical aperture of the objective is within the usual range for microdensitometry, i.e. below about $0 \cdot 4$; with the much higher numerical apertures commonly used for microfluorometry, non-linearity may develop at lower densities than those calculated above.

Concentration quenching

At high concentrations of fluorophore, there may be so much absorption of light from the excitation source that the intensity of excitation is significantly reduced in the inner region of the specimen; this results in a fall-off of observed fluorescence, known as concentration quenching.

Reabsorption of fluorescence

If the excitation (absorption) and emission curves of the fluorophore overlap, as they commonly do, some of the emitted light may be reabsorbed by the fluorophore. For this reason, it is important that the fluorescence emission be measured at a wavelength long enough to be beyond the effective limit of the absorption spectrum of the fluorescent substance. The possible complications due to reabsorption have been considered by several workers, notably Rigler (1966). A further factor, which does not appear to have been considered previously, is that if light is reabsorbed by the fluorophore, further fluorescence emission may ensue; and theoretically this process may repeat. If the amount absorbed is small, and/or if the quantum efficiency is low, the amount of what might be termed second-order or second-turn fluorescence will be relatively much smaller and may be ignored.

Table 4.1. **Relationship of extinction, absorption and transmission, calculated from formulae given in the text**

The limiting value of the ratio A/E is ln 10, which is 2·302 58 to six significant figures.

Extinction E	Absorption A	Transmission T	Ratio A/E
$1·0 \times 10^{-9}$	$2·302 \times 10^{-9}$	0·999 999 997 7	2·303
$1·0 \times 10^{-8}$	$2·302 \times 10^{-8}$	0·999 999 977	2·303
$1·0 \times 10^{-7}$	$2·302 \times 10^{-7}$	0·999 999 77	2·303
$1·0 \times 10^{-6}$	$2·302 \times 10^{-6}$	0·999 997 7	2·303
$1·0 \times 10^{-5}$	$2·302 \times 10^{-5}$	0·999 977	2·303
$1·0 \times 10^{-4}$	$2·302 \times 10^{-4}$	0·999 77	2·302
$1·0 \times 10^{-3}$	$2·299 \times 10^{-3}$	0·997 7	2·300
0·01	0·022 76	0·977 2	2·276
0·02	0·045 01	0·955 0	2·250
0·03	0·066 75	0·933 3	2·224
0·04	0·087 99	0·912 0	2·200
0·05	0·108 75	0·891 3	2·175
0·06	0·129 04	0·871 0	2·151
0·07	0·148 86	0·851 1	2·127
0·08	0·168 24	0·831 8	2·103
0·09	0·187 17	0·812 8	2·080
0·10	0·205 67	0·794 3	2·057
0·15	0·292 05	0·707 9	1·947
0·20	0·369 04	0·631 0	1·845
0·25	0·437 66	0·562 3	1·751
0·30	0·498 81	0·501 2	1·663
0·35	0·553 32	0·446 7	1·581
0·40	0·601 89	0·398 1	1·505
0·45	0·645 19	0·354 8	1·434
0·50	0·683 77	0·316 2	1·368
0·60	0·748 8	0·251 2	1·248
0·70	0·800 5	0·199 5	1·144
0·80	0·841 5	0·158 5	1·105
0·90	0·874 1	0·125 9	0·971
1·0	0·9	0·1	0·900
2·0	0·99	0·01	0·495
3·0	0·999	0·001	0·333
4·0	0·999 9	0·000 1	0·250
5·0	0·999 99	0·000 01	0·200
6·0	0·999 999	0·000 001	0·167
7·0	0·999 999 9	0·000 000 1	0·143
8·0	0·999 999 99	0·000 000 01	0·125
9·0	0·999 999 999	0·000 000 001	0·111
10·0	0·999 999 999 9	0·000 000 000 1	0·100

Distributional error

Distributional error, best known in microdensitometry, is due to non-uniform distribution in the measuring field of the substance being measured. At low concentrations of fluorophore, such that the intensity of fluorescence is linearly proportional to the concentration of the fluorophore, distributional error may be neglected; indeed this is one of the main advantages of microfluorometry.

Errors relating to the specimen

Fluorescence fading

It is characteristic of fluorescent preparations that they nearly always fade during irradiation. The rate of fading depends on the specimen, its environment, and the wavelength and intensity of irradiation. Some fluorophores are particularly liable to rapid fading (e.g. porphyrins) while others are relatively stable (such as NADH). After fading has taken place, some recovery may occur if the specimen is kept in the dark, preferably at low temperature.

Fading, otherwise known as photobleaching, is discussed in detail in Chapter 11. The following summarizes those aspects which are immediately relevant. Fading is essentially due to a photochemical reaction, induced by the light used for excitation. The absorption of light (prerequisite for fluorescence) entails the raising of molecules to the excited state, in which they tend to be more reactive than in the ground state. A small but significant proportion of the excited molecules, instead of fluorescing, undergo a photochemical reaction with the production of a different chemical species which may be non-fluorescent, or non-absorbent at the excitation wavelength.

Some fading of fluorescence is inevitable during microfluorometry. The rate of fading depends upon the individual fluorophore, its chemical environment, and the intensity and quantum energy of the irradiation. As shown by the shape of fluorescence fading curves, the proportion of excited molecules at risk of decomposition falls during irradiation, i.e. some molecules are more labile and 'fade' first. If the specimen contains a mixture of fluorescent substances, these may fade at different rates.

In quantitative microfluorometry, the steps required to minimize inaccuracies due to fading include:

1. Selection, where possible, of a resistant fluorophore (e.g. for the Feulgen reaction, selection of a resistant fluorochrome).

2. Selection of mounting media to reduce the likelihood of photochemical reactions.

3. Reduction of the amount of irradiation, by:
   (a) Localization of the specimen by phase-contrast or another method involving light of a low quantum energy.
   (b) Choice of an appropriate excitation wavelength without extraneous light of shorter wavelength.

(c) Minimization of the time lag between the commencement of irradiation and the completion of measurement, e.g. by using a photon-counting system triggered by the illumination shutter.

(d) The use of an efficient detector system (so that emitted light is not wasted).

(e) Use of a brief flash of excitation.

It is evident that the most important factor is minimization of the irradiation. This requires that the detection system should be as efficient as possible: a high-aperture monochromator or high-transmission filters, a sensitive photomultiplier and a photon-counting electronic system seem to be indicated. Another method is to measure the fluorescence at various known times after the commencement of irradiation, and extrapolate back to a theoretical intensity at zero time of irradiation. I have attempted this method, but without much success because the rate of fading is not constant (see Chapter 11).

Fluorophore not following the Beer–Lambert Law

There will be further non-linearity in the relationship between fluorescence intensity and fluorophore concentration if the Beer–Lambert Law is not obeyed; this may occur if the fluorophore changes its nature at different concentrations, or if the orientation of dichroic absorbers in it is not random. For example, ionization may be repressed and aggregation favoured at high concentrations. Metachromatic dyes such as Acridine Orange and Coriphosphine O are particularly subject to this problem.

Absorption of light by substances other than the fluorophore

Absorption of the light by some substance other than the fluorophore can lead to several possible complications. If the second substance is fluorescent, its emission will be added to that of the fluorophore of interest, probably resulting in over-estimation. A similar effect would result if the absorbing substance is itself non-fluorescent but transfers the absorbed energy to a fluorescent molecule (such as the fluorophore of interest, or another). Finally, if light is strongly absorbed by the unwanted absorber, an effect similar to concentration quenching is to be expected.

Autofluorescence

Tissue autofluorescence has the effect of increasing the background (see below). Autofluorescence of tissue is due mainly to NADH, flavins and proteins. Therefore, its intensity can be minimized by long-wavelength excitation. Its effects can also be minimized by illumination at a wavelength corresponding to the excitation maximum of the specific fluorophore, which is thereby favoured in comparison with other fluorescent substances present unless they happen to have similar excitation maxima. It may be necessary to subtract a 'blank' measurement of tissue autofluorescence measured from a control area (Ritzén, 1967). Sick & Rosenthal (1989) reported difficulty in living cells due to autofluorescence of NADH, of which a significant proportion (10%) was labile due to experimental electrical stimulation and anoxia.

Unsuitable fluorochrome

A fluorochrome used for microfluorometry can introduce errors due to photobleaching (fading under irradiation), failure to obey the Beer–Lambert law, changes in colour due to metachromasia, and non-specific staining. The criteria for choosing a fluorochrome are discussed in Chapter 5.

Instrumental errors

Instability of the light source

A stable light source is essential in instruments not incorporating a reference channel, and is advantageous in any case as it is more convenient for the operator if the light intensity remains reasonably constant. After the initial warming-up period (30 min or so), short-term variations in the light from mercury or xenon arc lamps are due to wandering of the arc (producing flickering) and to change in lamp current due to variations in supply voltage. The intensity of the arc can be stabilized by regulating the current flowing through it. This requires a direct-current source, with appropriate control circuitry. Alternatively, the conventional power unit of the lamp can be fed from stabilized AC mains, e.g. by placing a regulating transformer between the mains supply and the conventional power unit of the lamp. This latter may not be entirely satisfactory, because the conditions of operation of the regulating transformer are far from optimal for two reasons. The transformer has to have a high rating to cope with the starting current, while the operating current is below that for which the transformer was designed; and the inductive load presented by the starting device results in a poor power factor, again rendering the regulation less effective. The power factor may be corrected by placing an appropriate capacitor in parallel with the starting device (Rost & Pearse, 1971).

Instability of arc position gives rise to flickering and sudden changes in the level of intensity as the arc skips from one part of the electrodes to another. The shape of the electrodes is designed to minimize wandering of the arc. In general, xenon arc lamps are more stable in this respect than mercury arc lamps. It is important that, when a new arc lamp is switched on for the first time, it be kept running for a substantial period of time (at least 24 h) for the arc to settle down. The Farrand MSP (see Chapter 16) had a magnetic stabilizer for the arc, to hold it steady.

Even if a reference channel is employed, a stable arc is still required so that the arc does not move out of alignment with the optical axis. If the reference channel is not exactly analogous with the main channel in position and aperture, flickering of the lamp may affect the two channels differently – e.g. if the arc image moves partly outside the aperture of the main illumination system while still shining fully into the reference channel. The procedure for checking the stability of the light source was described in Chapter 3.

Homogeneity of illumination

For homogeneous illumination of the object, the object plane must be illuminated by light which is of uniform intensity over the entire field, and, if the light is less than perfectly monochromatic and therefore consists of a mixture of wavelengths, light of all wavelengths must be uniformly distributed over the field. Tests for homogeneity of illumination are described in Chapter 3.

The first of these requirements (uniform intensity) rules out the possibility of imaging the light source in the object plane (so-called 'critical' or source-focussed illumination). The second requirement (uniform distribution of all wavelengths present) is not a problem if a filter is used rather than a monochromator (see Chapter 9). It is best to avoid imaging the surface of any field lens or mirror exactly into the object plane, since these surfaces may support particles of dust which will appear in the field.

Errors in the reference channel

A reference channel to compensate for variations in intensity of illumination caused by instability of the light source is more commonly used in microspectrofluorometers, and is therefore discussed in Chapter 9.

The direction of illumination

Three main possibilities exist for illumination: through a substage bright-field condenser, through a substage dark-field condenser, and through the objective (epi-illumination). The direction of illumination in relation to the direction of measurement affects the relationship between concentration of the fluorophore and measured fluorescence intensity (Fig. 1.4). Udenfriend (1962) pointed out in relation to cuvette fluorometry that, unless the specimen is illuminated from the same side as fluorescence is measured, at high fluorophore concentrations the apparent fluorescence intensity will be low due to reabsorption of emitted light by the specimen itself. Rigler (1966) discussed the applications of this principle to microspectrofluorometry, and concluded that epi-illumination was to be preferred because of greatest linearity of the fluorescence intensity/concentration relationship and least distortion of the spectral distribution of emitted radiation by reabsorption.

Background

All measurements of fluorescence intensity include a certain amount of unwanted 'background', which is due to the following:

1. Photomultiplier dark current and electrical noise in the measuring system.
2. Stray light of the emission wavelength, derived from the excitation source.
3. Stray light due to light leakage into the measuring system.
4. Stray light due to autofluorescence of the optical system.
5. Autofluorescence of the tissue and mountant.

The effects of (1) and (3) can be largely eliminated by chopping the light beam and using an AC amplifying system, as in the instrument of Caspersson, Lomakka & Rigler (1965). Otherwise, there are two basic measures which can be taken to minimize the background or its effects: the actual background can be reduced as much as possible, and a measurement of the background can be made and its value subtracted from the observed intensity to obtain a corrected measurement. Neither of these procedures is quite as simple as one might suppose.

Photomultiplier dark current is minimized mainly by selection of a suitable photomultiplier, keeping it dry, and cooling it. Cooling with solid carbon dioxide is sufficient to reduce the dark count from about 5000 to about 100 pulses per second or less. Keeping the photomultiplier dry is even more important; and at least care must be taken to avoid condensation of moisture on the photomultiplier window and on the high-tension circuitry. The signal/dark current ratio can be optimized by selection of the most suitable anode voltage for the particular photomultiplier tube, and (in photon-counting systems) by selection of optimal threshold for counting (see Chapter 12).

Stray light at the emission wavelength due to insufficient monochromation of the illumination is awkward to deal with. Such light is by definition shone onto the specimen by the illumination system, and some may be reflected back by the specimen and by optical components, and measured together with the emission. Techniques for minimizing this source of error are described in Chapter 3. A similar problem arises from light at the excitation wavelength not rejected by the emission filter. A technique for assessing the relative contributions of fluorescence and reflection to the measurement is to compare measurements made with the barrier filter (1) in its usual position, and (2) in the illumination pathway before the epi-illuminator (Galassi, 1990). Fluorescence should be very much reduced in the latter case, while reflected light is relatively unaffected.

Light leaks, i.e. stray light from the ambient illumination of the room may enter the instrument. So long as it remains constant, the effect will be to increase the apparent dark current; however, this form of stray light is subject to sudden variations due to movement of the operator's shadow or the opening of a door. Because of the extreme sensitivity of the photomultiplier system, great precautions have to be taken against light leaks and to minimize stray light. Some suggestions for checking for light leaks are given in Chapter 3. My own practice in most cases has been to use an empty field as the control area; this at least compensates adequately for stray light entering the system or reflecting from the objective (epi-illumination) but occasionally gives odd results probably due to the fact that (assuming epi-illumination) tissue in the measuring field reduces stray light reflected from the undersurface of the object slide or otherwise transmitted from below.

The effect of a given level of background on the accuracy of a measurement can be calculated as follows. The signal (S) is equal to the difference between the observed total photon count (N) and the background count (B). The variance of the difference is equal to the sum of the variances, so that the standard deviation (σ) of the signal count is given by:

$$\sigma = \sqrt{(N+B)}$$
$$= \sqrt{(S+B)+B}$$
$$= \sqrt{(S+2B)}$$

Table 4.2. **Total count (N) required for a given signal-to-background ratio (S/B) and for a given precision (E) expressed as the ratio (standard deviation of S)/S**

S/B	N		
	$E=1\%$	$E=0.3\%$	$E=10\%$
0·1	230 000	23 000	2 300
0·5	150 000	15 000	1 500
1	60 000	6 000	600
2	30 000	3 000	300
3	22 000	2 200	220
4	19 000	1 900	190
5	17 000	1 700	170
10	13 000	1 300	130
20	12 000	1 200	120
100	10 000	1 000	100
∞	10 000	1 000	100

The relative error is given by:

$$\frac{\sigma}{S} = \sqrt{\frac{S+2B}{S^2}}$$

The required total photon count (N) to obtain a specified accuracy (σ/S) for various ratios (f) of background to signal may be calculated from the formula:

$$N = \frac{(1+f)(1+2f)}{(\sigma/S)^2}$$

The total counts required for various values of signal to background to give the specified accuracies are given in Table 4.2.

Non-linearity of photometer

Linearity of the photometer can be checked in two ways: either by measuring a series of standard objects of known relative fluorescence, e.g. using the capillary tubes of Sernetz & Thaer (1970, 1973), or by measuring from a constant object while reducing the illumination by known amounts with neutral-density filters. The latter method is much more convenient, provided that the standard object does not fade. A liquid solution of fluorochrome is suitable. Neutral density filters of accurately known transmission are made by Carl Zeiss (Oberkochen, West Germany).

Lost counts are a source of non-linearity in photon-counting systems. If a counting device which is counting random pulses has a finite resolving time not all the pulses will be counted, due to some pulses arriving too soon after a previous pulse. If the counter is insensitive during a constant time p, and assuming that the input pulses arrive at random intervals (according to Poisson statistics), if the observed count is N per second the best estimate (Ruark & Brammer, 1937) of the true count is $N/(1 - Np)$. Depending on the pulse-processing circuitry, the deadtime may be extendible: incoming pulses separated by less than the deadtime p interact and give rise to a single, longer output pulse; this results in the loss of all pulses separated by a time less than the deadtime. For further discussion, see Cheng, Crozier & Egerton (1987). If a photon-counting system is used, non-linearity due to lost counts at high count rates together with all other causes of non-linearity can be lumped together. From the shape of the curve, an apparent deadtime can be calculated; an appropriate correction can be calculated to cover both deadtime and any other causes of non-linearity.

Unstable photometer

The photomultiplier anode current is strongly dependent on the anode voltage. Therefore, a highly stable high-tension (EHT) power pack is essential for reliable current measurements; if photon counting is used the stability of the power supply is less critical, because the number of pulses is counted rather than the amplitude of the current; hence, only the background (dark current) is affected by changes in voltage.

STANDARDIZATION

Unlike microdensitometry, in which an absolute determination of optical density is made, microfluorometry is strictly comparative and relies on a standard for its calibration. Whereas in absorptiometry, measurements are made of the intensities of the incident and transmitted light, and the proportion absorbed calculated, in fluorometry the intensities of the incident and fluorescent light are to be measured and the ratio calculated. The problem lies in the measurement of the incident light, which in a microdensitometer is determined by measurement of the empty field (without the specimen), but cannot be measured directly in microfluorometry.

Microfluorometry has therefore come to rely exclusively on the comparison of the fluorescence of the object to be measured with that of a standard preparation, under stable conditions. The requirements for accurate standards have been discussed by, among others, Ritzén (1967), Sernetz & Thaer (1970, 1973), and Jongsma, Hijmans & Ploem (1971); the matter is also dealt with further, in a different context, in Chapter 9. To ensure proper comparability of standard and test objects, the following criteria (largely due to Ritzén, 1967; see also Lichtensteiger, 1970) may be laid down in respect of the standard:

1. Its size should be similar to that of the objects being measured, so that conditions of microscopy can be identical.
2. The standard must have the same fluorescence excitation and emission spectra as the object otherwise it will respond differently to changes in the excitation. This means that it should be made of the same chemical substance, in the same chemical environment. Unfortunately the fluorescence characteristics of a fluorochrome bound to a larger molecule (such as a protein) are not quite the same as those of the unbound fluorochrome, so that a solution of a fluorochrome is not an

ideal standard for bound fluorochrome. A protein-bound standard was described by Goldman & Carver (1961).

3. The standard is required to be constant, not fading during the measurement process and preferably also constant during storage.

4. The standard must be reproducible.

5. The standard must contain an accurately known amount of fluorophore.

6. If the concentration of fluorophore is high, the standard must be thin enough to ensure sufficiently uniform excitation throughout the entire layer, and to minimize the reabsorption of fluorescence emitted. This is particularly important if dia-illumination is used.

7. Ideally, the standard should be easy to obtain or make.

Several types of fluorescence standard have been described in respect of microfluorometric mass determinations in cytochemistry; the best at present (for most purposes) is the capillary tube method of Sernetz & Thaer (1970, 1973); other standards include uranyl glass, phosphor particles, fluorochrome solutions in microdroplets or a cuvette, and fluorochromed beads of carbohydrate or latex. These are described below.

Fluorochrome solution

For standardization of measurements of a fluorochrome, the most easily available and reproducible standard is generally a solution of the fluorochrome in question. The solution must be buffered to an appropriate pH, and (ideally) bound onto a molecule to represent the intended substrate. A standard fluorescein solution was described by Jongsma et al. (1971; see Appendix 2).

A cuvette with fluorochrome solution has been used, but this is generally unsatisfactory and has been largely superseded by capillary tubes and microdroplets as described below. Using epi-illumination, the objective can be dipped into a dish of fluorochrome solution using the aperture-defined volume principle.

Sernetz–Thaer capillary tubes

The capillary tube method of Sernetz & Thaer (1970, 1973) is based on capillary tubes of quartz-glass or glass with an inside diameter of 6–8 μm, an external diameter of about 30 μm, and about 50 mm long. These tubes are filled with a fluorochrome solution of known concentration, and the fluorescence of a measured length of capillary tube is measured.

Sernetz & Thaer (1970, 1973) used capillary tubes produced at the Battelle Laboratories in Frankfurt according to a procedure described by Nixdorf (1967). Up to 20 capillary tubes are embedded side by side between a normal microscope slide and a coverglass in a medium of refractive index equal to that of the glass. Normal glass capillaries may be embedded in DPX, Entellan, or other non-fluorescent embedding medium, while quartz-glass capillaries may be embedded in glycerol. If glycerol is used as the embedding medium, the capillaries have to be fixed, e.g. by a narrow strip of silicone grease, in order to prevent displacement of the capillaries between slide and coverglass during the embedding process. If the excitation wavelength is to be shorter than 360 nm, care must be take to use ultraviolet-transparent materials where required: quartz-glass capillaries, and either a quartz coverslip (for epi-illumination) or a quartz slide (for dia-illumination). A quartz coverslip is much cheaper than a quartz slide, another advantage of epi-illumination.

While the tubes are empty, the vertical inside diameter of the embedded capillary tubes is measured microinterferometrically, by the displacement of interference fringes. The diameter is calculated from the expression:

$$d = \Gamma / \Delta n$$

where d is the inside diameter of the capillary tube (μm), Γ is the path difference (μm) and Δn is the difference in refractive index between capillary glass and capillary lumen. If an interference microscope is not available, I imagine that the size of the lumen could be determined by weighing the tube before and after filling with water; this presupposes that the tube be loose, not yet mounted in a slide.

To fill the embedded capillary tubes, a droplet of fluorochrome solution is sucked from the edge of a coverglass into each of the capillary tubes by gentle sucking at the other end of the tube. The penetration of the solution into the capillary lumen is easily observed. After filling, the capillary tube ends projecting beyond the coverglass are cut off and the ends sealed with a mounting medium such as DPX or Entellan. It is possible to fill individual tubes with different fluorochromes or with different concentrations of fluorochrome.

The practical procedure for using the Sernetz–Thaer tubes is generally as follows. The tube is obtained, mounted in a slide, and its diameter measured as described above. The tube is filled with standard fluorochrome solution of known concentration. The slide is placed on the stage of the microflurometer, and a portion of a capillary tube located centrally in the measuring field. The fluorescence from this portion of the tube is measured (see Fig. 4.1). The length of tube in the field is measured with an eyepiece graticule, or from the known diameter of a fixed measuring field (which can be measured with a stage micrometer; see Fig. 4.2). The absolute amount of fluorochrome in the measuring field may be calculated from the formula:

$$W = \pi d^2 sc$$

where W is the mass of fluorochrome, d is the diameter of the capillary tube, s the length measured, and c the concentration of fluorochrome in the solution, all in compatible units. As regards the units, if d and s are expressed in millimetres, and c in milligrams per microlitre (which is the same as grams per millilitre), then W will be expressed in milligrams.

Jongsma *et al.* (1971) compared the use of standard fluorescein solutions in capillary tubes with uranyl glass standards, in relation to immunofluorescence, and found the former preferable. This is true particularly of transmitted-light excitation, but also of incident-light excitation.

Compared with the microdroplet method the capillary method offers substantial advantages. With the capillary standard, the excited section with a length of about 50 μm is very small in comparison with the total capillary length. During fluorescence excitation the fluorochrome solution of the excited section is continuously exchanged with non-irradiated solution, so that photodecomposition does not become effective. Therefore, an almost constant standard value is obtained even during lengthy excitation, whereas the microdroplets show a more or less strong irreversible decrease in the fluorescence signal during the measurement depending on the radiation intensity even with the relatively photoresistant fluorescein. Whereas a microdroplet can be measured only once because of fading under irradiation, in capillary tubes the fluorescence intensities were found to remain constant over two months and many measurements, and may be expected to continue to do so provided the fluorescent standard solution itself does not change its properties responsible for fluorescence emission.

Because of the constancy of their properties the capillary standards may also be used for testing the homogeneous illumination of the illuminated area in the object plane.

Microdroplets

The microdroplet method (Rotman, 1961; Rigler, 1966; Ploem, 1970; Jongsma *et al.*, 1971) is based on the measurement of the fluorescence intensity of single droplets with diameters up to about 20 μm sprayed onto a silicone oil layer (Ploem, 1969). The

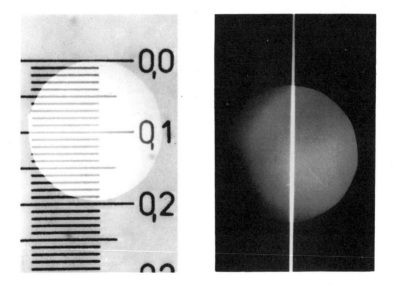

Fig. 4.1. (*Left*) Measurement from Sernetz–Thaer capillary tube. The picture shows a fluorescent capillary tube, containing a solution of fluorescein isothiocyanate (FITC) which is excited by epi-illumination. For the purposes of the photograph only, an image of the measuring field diaphragm has been made to appear in the field, to show what is happening.

Fig. 4.2. (*Right*) Measurement of the diameter of a measuring field, using a stage micrometer. The measuring field, limited by an iris diaphragm, is defined by the central bright area.

measured fluorescence intensity can be related to the mass of fluorochrome in the droplets. The fluorochrome content of single droplets is calculated from their measured lateral diameters and the known concentration of fluorochrome in the solution. In practice, several droplets should be measured, and an average calibration factor determined. Because of fading, an individual droplet can be measured only once.

This method meets all requirements in respect of the standard's position in the object plane and of the standard's similarity to microscopic objects. It also permits the relevant fluorochrome or substances stained with fluorochromes to be used as the standard.

Before the development of the microdroplet method, there was no methical basis for the direct correlation of the standard value and thus of the measured intensities of the fluorescent objects to absolute amounts of the fluorescent substance, apart from the use of biological standards stained under exactly the same conditions as the objects to be measured.

There are two main sources of error with the microdroplet method. Firstly, volume determination in the microdroplet method requires the exact maintenance of the spherical shape of the droplets. Apart from this condition, the experimental error of the diameter determination affects the volume calculation of the droplet more severely than that of the capillary, since in the former case the measured diameter is cubed and in the latter squared (the error in determining the length of the measured capillary section is relatively low). Secondly, with the microdroplet method the entire volume of the sphere is excited and is subject to photobleaching (the degree of which depends on the kind of fluorochrome, the time of radiation and the radiation intensity applied); whereas, with the capillary tube method (described below), only a part of the solution is irradiated.

The microdroplet method can be adapted to standardization of induced fluorescence (e.g. formaldehyde-induced fluorescence (FIF)) by measuring out droplets of known volume and drying the droplet before exposure or vapour treatment. For FIF, a protein must generally be incorporated (Ritzén, 1967; Lichtensteiger, 1970); other substances may be usable, such as polyvinylpyrrolidone (PVP; Rost & Ewen, 1971).

Fluorescent (uranyl) glass

Fluorescent glass contains fluorescent inorganic ions, usually uranyl or rare earth ions. Uranyl glass, for instance Schott type 6617, does not change its fluorescence significantly in the short term even under strong irradiation by the exciting light. A plate of glass can be placed in the object plane (Eder, 1966), between the illumination field diaphragm and condenser (Rigler, 1966), or immediately below the objective (assuming epi-illumination; Ruch, 1970). Because of their dimensions, plates cannot be measured under the same conditions of microscopy as the objects to be investigated; this problem is solved by using microspheres or thin fibres of fluorescent glass (van der Ploeg et al., 1977b).

A standard fixed in a magnetic disc, as used by Zeiss (Oberkochen) for reflectivity measurements, is very practical (Ruch, 1970). The disc can quickly be attached to the microscope objective. Since the fluorescing plate is held in a fixed plane of the microscope it gives repeatable values, provided that the field and illumination diaphragms are kept constant.

Apart from the focus problem, the layers of the glass plate above and below the focussed level make different contributions to the total emission and to the emitted fluorescence light taken up by the objective. These different contributions depend critically on the illumination aperture and the size of the illuminated area. Of course, in the case of dia-illumination, this problem can be solved by placing the uranyl glass standard anywhere between the illumination-field diaphragm and the condenser (Rigler, 1966). This latter arrangement has the further advantage that inhomogeneities in the standard are averaged out and that the radiating flux per unit area of the standard plate is low compared with the position of the standard in the object plane. These advantages, however, do not compensate for the fact that the object plane on the one hand and any positions between field diaphragm and substage condenser on the other are not interchangeable in respect of fluorescence excitation and emission, and that also because of the required similarity to the fluorescing objects the standard has to be situated in the object plane. Apart from this general disadvantage it is not possible to place the fluorescence standard in any position of the exciting beam between dichromatic mirror and light source in the case of incident-light excitation, usually considered as the most favourable optical arrangement for microfluorometry. These objections also apply to a position of the fluorescence standard between objective and image plane below the barrier filter. This position has been discussed by Prenna & Bianchi (1964) for transmitted-light excitation.

Besides glass, solid-block fluorescence standards are available from various sources, either sold as such, e.g. the Perkin-Elmer standard (Perkin-Elmer, Norwalk, CT, USA) or as scintillators (Koch-Light, Haverhill, Suffolk CB9 8PU, UK).

Phosphor crystals

Crystals of phosphors can be mounted on a slide and embedded in DPX or other non-fluorescent medium under a coverslip. Their luminescence has been used as standard (Goldman, 1967; Pittman et al., 1967; see also West & Golden, 1976). They are small enough to be used under conditions of microscopy similar to those of the object to be measured, and are reasonably stable under irradiation. On the other hand, their luminescence characteristics are different from those of fluorochromes likely to be used in microfluorometry. Each crystal is different and must be individually identified; time is lost searching each time for the marked crystals. Also, there is difficulty in achieving reproducible focussing with high-power objectives.

Fluorochromed carbohydrate beads

Microscopical beads of a carbohydrate material were first introduced for cytochemical models by van Duijn & van der Ploeg (1970), who emulsified a viscose solution. Subsequent studies have been based on commercially available beads of dextran (Sephadex) or agarose (Sepharose). Aminoethyl-Sephadex beads, labelled with either Tetramethyl Rhodamine or fluorescein isothiocyanate (FITC), were introduced by Haaijman & van Dalen (1974; see also Haaijman & Slingerland-Teunissen, 1978). As in the microdroplet method, the diameter of the beads is measured with an eyepiece micrometer. For a review, see van der Ploeg & Duijndam (1986).

Defined-substrate sphere system (DASS)

The use of Sephadex beads can be extended to the conjugation of specific proteins to beads, which may then be stained with immunofluorescent reagents. This method has been investigated by Knapp & Ploem (1974).

Biological standards

The artificial standards described above permit only relative calibrations to be made. Biological objects with a known content which are stained under exactly the same conditions as the objects to be measured give an absolute calibration, to the extent that the true value is known. Biological standards which have been used for the calibration of DNA are bull spermatozoa (Ruch & Bosshard, 1963; Ruch, 1966b), and the bacterium *Pediococcus damnosus* (Cerevisiae) (Lawrence & Possingham, 1986b).

Comparison with microdensitometry

An excellent check on any quantitative method is the use of an independent method. One step is to check or calibrate microfluorometry by microdensitometry of the same preparation. Direct comparison of microdensitometry and microfluorometry on the same specimens may assist in calibration as well as validation of the overall technique. Such comparisons of microfluorometry with microdensitometry were made in respect of quantification of nuclear DNA by Böhm, Sprenger & Sandritter (1970, 1973), Schnedl *et al.* (1977), Geber & Hasibeder (1980), Tanke & van Ingen (1980), and Broekaert *et al.* (1986); and in respect of quantitative immunocytochemistry by Smith, Redick & Baron (1983). For quantification of nuclear DNA, Böhm *et al.* (1970, 1973) compared microfluorometry with microdensitometry, using Feulgen staining with an Acriflavine–Schiff-type reagent. These days, using green-light excitation, the same comparison is possible with the standard Schiff reagent (see Prenna, Leiva & Mazzini, 1974; Tanke & van Ingen, 1980). A second step is to use an independent cytochemical method as well; thus Geber & Hasibeder (1980), also quantifying DNA, compared fluorometry using DAPI with microdensitometry of the Feulgen reaction (see Chapter 5). For quantitative immunocytochemistry, Smith *et al.* (1983), studying the distribution of a reductase in liver, compared microdensitometric quantification of unlabelled antibody peroxidase–antiperoxidase staining and microfluorometric quantification of indirect fluorescent antibody staining.

DATA PROCESSING

It has already been indicated that, for each measurement, it is necessary to calculate the results, x, according to the formula:

$$x = \frac{(\text{nominal value of standard}) \times [(\text{reading of test}) - (\text{reading of test blank})]}{(\text{reading of standard}) - (\text{reading of standard blank})}$$

If a reference channel is used, each of the readings in the above formula must be divided by the simultaneous reading of the reference channel (less the value of the reference channel's blank, which is ordinarily a negligible dark current). The acquisition of data and the calculations involved are most easily accomplished by an on-line computer, which can also calculate means and standard deviations of results: see, for example, Rundquist (1981) and Rundquist & Enerbäck (1985).

Statistical analysis of results is best planned in advance, before making the measurements. Useful basic texts on statistics include, at an elementary level, Byrkit (1987), Campbell (1989), Goldman & Weinberg (1985), Lumsden (1969) and Rowntree (1981); and at a more advanced level, Kreyszig (1970), Steel & Torrie (1980), and Walpole & Myers (1978). Other texts include those of Armitage (1971), Derman & Klein (1959), Dixon & Massey (1968), and Freeman (1963).

5

Applications of microfluorometry

The development of microfluorometry was preceded by that of microdensitometry. The main stimuli for the development of microfluorometry were the need for quantification of fluorescent cytochemical procedures (such as immunocytochemistry and the formaldehyde-induced fluorescence of neurotransmitter amines), and the fact that individual microfluorometric measurements can be made much more quickly than corresponding measurements by scanning densitometry and therefore microfluorometry greatly facilitated the repetitive measurements required for the study of substantial cell populations. Microfluorometry therefore came to be most widely used for the measurement of nuclear deoxyribonucleic acid (DNA) by the Feulgen method (using a fluorescent Schiff-type reagent). For the same reason, classical microfluorometry has now been substantially replaced by flow cytofluorometry (see Chapter 14), where possible.

The only previous overall reviews of applications of microfluorometry appear to be those of Ruch (1964, 1966b, 1970). On more specific topics, there have been reviews on the quantification of DNA using the fluorescent Feulgen technique (Prenna, 1968), immunofluorescence (Goldman & Carver, 1961; Ploem, 1970; Sternberger, 1986) and neurotransmitter amines using formaldehyde-induced fluorescence (Ritzén, 1967; Lichtensteiger, 1970; Jonsson, 1971). Some examples of applications of microfluorometry are listed in Table 5.1.

Microfluorometry: when to use it

Microfluorometry is obviously appropriate and indeed necessary for the quantification of autofluorescence and cytochemical reactions with a fluorescent end-point, such as formaldehyde-induced fluorescence. In other cases it may be possible to choose between microfluorometry and microdensitometry; in such case it may be helpful to consider reports of workers who have applied both to the same problem Such comparisons of microfluorometry with microdensitometry were made in respect of quantification of nuclear DNA by Böhm, Sprenger & Sandritter (1970, 1973), Schnedl et al. (1977), Geber & Hasibeder (1980), Tanke & van Ingen (1980), and Broekaert et al. (1986); and in respect of quantitative immunocytochemistry by Smith, Redick & Baron (1983).

Table 5.1. **Applications of microfluorometry**

This list is not exhaustive, particularly in respect of recently developed probes for ionic concentration.

Substance	'Stain'	Authors	Year
Autofluorescent pigments			
Lipofuscin	Autofluorescence	Dowson & Harris	1981
	Autofluorescence	Collins & Thaw	1983
Nucleic acids			
DNA (Feulgen)	Auramine O	Bosshard	1964
	BAO	Ruch	1966b
	?	Bahr & Wied	1966
	Acriflavine	Böhm *et al.*	1970
	BAO	Ruch	1970
	Acriflavine, Acridine Yellow Coriphosphine	Böhm *et al.*	1973
	Pararosaniline	Fujita	1973
	Pararosaniline	Prenna, Leiva & Mazzini	1974
	BAO	Kemp, Doyle & Anderson	1979
	Acriflavine	Tsuchihashi *et al.*	1979
	Acriflavine	Tanke & van Ingen	1980
	Auramine O	Číhalíková, Doležel & Novák	1985
	Pararosaniline	Broekart *et al.*	1986
	Pararosaniline	Giangaspero *et al.*	1987
	Acriflavine	Wouters *et al.*	1987
DNA	Acridine Orange	Donáth	1963
	Bisbenzimide	Latt	1973
	Mithramycin	Johannisson & Thorell	1977
	DIPI	Schnedl *et al.*	1977
	Ethidium bromide	Inoki, Osaki & Faruya	1979
	Bisbenzimide	Cowell & Franks	1980
	DAPI	Geber & Hasibeder	1980
	DAPI, Mithramycin	Coleman, Maguire & Coleman	1981
	Bisbenzimide	Gordon & Parker	1981
	DAPI, Chromomycin A_3	Leemann & Ruch	1982
	Acridine Orange	Hemstreet *et al.*	1983
	DAPI	Hamada & Fujita	1983
	DAPI	Coleman	1984
	DAPI	Goff & Coleman	1984
	Acridine Orange	Ostling & Johanson	1984
	Acridine Orange	Rosen & Mercer	1985
	DAPI	Levi *et al.*	1986
	DAPI	Lawrence & Possingham	1986a,b
RNA	Acridine Orange	Thaer & Becker	1975
Chromatin	Morin	Cowden & Curtis	1981
Proteins and amino acids			
Non-histone nuclear proteins	Fluorescein mercuric acetate	Cowden & Curtis	1973
Non-histone nuclear proteins	Brilliant Sulphoflavine	Ruch	1970, 1973
Chromosomal proteins	DANS, Fluorescamine	Lee & Bahr	1983
Nuclear proteins	BSF, AO, CPM	Cowden & Curtis	1981

Table 5.1. *(cont.)*

Substance	'Stain'	Authors	Year
Arginine	Ninhydrin	Rosselet	1967
Lysine	DANS	Rosselet & Ruch	1968
Carbohydrates			
Carbohydrates	PAS	Prenna	1968
	PAS	Changaris, Combs & Severs	1977
Glycogen	PAS	Gahrton & Yataganas	1976
Glycogen	PAS	Tsuchihashi *et al*	1979
Heparin	Berberine	Enerbäck	1974
	Berberine	Gustavsson & Enerbäck	1978
	Berberine	Gustavsson	1980
Immunofluorescence and lectin binding, etc.			
Immunofluorescence	FITC	Goldman	1967
		Nairn *et al.*	1969
		Taylor & Heimer	1975
		Wahren	1978
	FITC	Fliermans & Hazen	1980
	FITC	Smith *et al.*	1983
	FITC	Huitfeldt *et al.*	1987
Non-specific staining	Immunofluorescence	Pittman *et al.*	1967
Sialoglycoconjugates	Lectins	Barni, Gerzelli & Novelli	1984
ABH isoantigens	Immunofl./lectins	Borgström & Wahren	1985
Probes			
Axonal transport	Primuline	Enerbäck, Kristensson & Olsson	1980
Pinocytosis	Dextran uptake	Berlin & Oliver	1980
Mitochondria	DASPMI	Horster, Woilson & Gundlach	1983
Uptake	Haematoporphyrin	Schneckenburger *et al.*	1987
	Haematoporphyrin	Schneckenburger & Wustrow	1988
pH	Fluorescein thiocarbamyl ovalbumin	Heiple & Taylor	1982
pO_2	Pyrene	Podgorski *et al.*	1981
Ca^{2+}	Various	Fabiato	1982
	Fura-2	Kassotis *et al.*	1987
	Quin-2	Uematsu *et al.*	1988
	Indo-1	Sick & Rosenthal	1989
Neurotransmitter amines, etc.			
Catecholamines	FIF	Ritzén	1967
	FIF	Lichtensteiger	1970
	FIF	Jonsson	1971
	FIF	Bacopoulos *et al.*	1975
(Dopamine)	FIF	Einarsson, Hallman & Jonsson	1975
	FIF	Löfström, Jonsson & Fuxe	1976
		Löfström *et al.*	1976
(Dopamine)	FIF	Agnati *et al.*	1979
	FIF	Partanen, Hervonen & Alho	1980
		Partanen, Hervonen & Rapaport	1982

Table 5.1. *(cont.)*

Substance	'Stain'	Authors	Year
(Dopamine)	Glyoxylic acid	Redgrave & Mitchell	1982
	FIF	Andersson *et al.*	1983
	FIF	Andersson & Eneroth	1985
Serotonin	FIF	Tilders *et al.*	1974
	FIF	Enerbäck & Jarlstedt	1975
	FIF	Jonsson *et al.*	1975
	FIF	Geyer *et al.*	1978
	FIF	Enerbäck & Gustavsson	1977
	FIF	Gustavsson	1980
	FIF	Alho	1984

Requirements for a fluorochrome for fluorometry

Any fluorochrome used for microfluorometry should be chosen having the requirements of quantification specifically in mind, in order to obtain optimal specificity, sensitivity, linearity, and signal-to-noise ratio. In studies involving the use of a fluorochrome there are usually several possible fluorochromes from which to choose. The following criteria should be considered when selecting a fluorochrome. The fluorochrome should, as far as possible:

1. Stain specifically.
2. Obey the Beer–Lambert law (see Chapter 4).
3. Have no tendency to metachromasia (change of colour due to polymerization at sites of high local concentration).
4. Have a high quantum efficiency.
5. Be resistant to photobleaching (fading under irradiation, see Chapter 11).
6. Have widely separated excitation and emission spectra (to reduce reabsorption, see Chapter 4).
7. Be excited at a wavelength substantially longer than that giving rise to autofluorescence (this may conflict with item (6)).

Using a fluorochrome as a dye rather than as a label, it is important when reporting results to name the fluorochrome as precisely as possible, citing the *Colour Index* number (Anon., 1971) if possible. For example, there appears to have been some confusion between Sulphaflavine and Brilliant Sulphoflavine FF (Rost, 1990).

Applications

Nucleic acids
In biology, microfluorometry has most commonly been applied to the measurement of nuclear DNA. Fluorescent staining of nucleic acids is discussed in *Fluorescence*

microscopy (Rost, 1990). Earlier reviews were written by Ruch (1966b, 1973), Böhm & Sprenger (1968) and Böhm *et al.* (1973). For microfluorometry the most valuable methods have been the Feulgen method with conventional (Pararosaniline) or Acriflavine–Schiff (for current method see Tanke & van Ingen, 1980), and various intercalating dyes (see Curtis & Cowden, 1983). Acridine Orange, a metachromatic dye, has been used by Hemstreet *et al.* (1983) who showed that under their staining conditions, the green (orthochromatic) component of the fluorescence was stoichiometrically related to the DNA content of nuclei; however, I would not recommend this method, considering the availability of probably more reliable methods; and if the method is used, probably a narrow-band green filter should be used for selecting the emission band. Microfluorometry has substantially replaced microdensitometry for DNA measurement, because of greater rapidity of measurement; and for the same reason has to a considerable extent itself been superseded by flow cytometry, using the same fluorescent staining techniques (see Chapter 14).

Ribonucleic acid (RNA) has been much less often quantified. Thaer & Becker (1975) quantified RNA in reticulocytes (young erythrocytes) using Acridine Orange (DNA being absent from human erythrocytes).

The Schiff reaction
Quantification of the Schiff reaction has been applied mainly to the Feulgen method for DNA (discussed below) but also to the periodic-acid–Schiff method, for studying glycogen and other cytoplasmic 1,2-glycols (Prenna, 1968; Gahrton & Yataganas, 1976; Changaris *et al.*, 1977; Tsuchihashi *et al.*, 1979). This method has also been combined with Acriflavine–Feulgen staining for DNA (Tsuchihashi *et al.*, 1979).

Immunofluorescence
The quantification of immunocytochemistry has been briefly reviewed by Sternberger (1986) and Ploem (1970). Quantitative immunocytochemistry has a lot of difficulties, of which probably the least is the microfluorometry. One of the main problems is standardization; see Chapter 4 and Wick, Baudner & Herzog (1978), Haaijman & van Dalen (1974), Taylor & Heimer (1975), and Jongsma, Hijmans & Ploem (1971). Recent studies applying quantitative immunofluorescence to cytochemistry have tended to use flow cytofluorometry (see Chapter 14).

Neurotransmitter amines: formaldehyde-induced fluorescence
The quantification of formaldehyde-induced fluorescence of neurotransmitter amines was first investigated by Eränkö & Räisänen (1961) and Ritzén (1967). This is a topic where microfluorometry offers the only hope of cytochemical quantification, since the corresponding non-fluorescent methods (Eränkö & Räisänen, 1961) are not sensitive enough for use in nervous tissue. The review by Ritzén (1967) is now classic; more recent ones are those by Lichtensteiger (1970) and Jonsson (1971). Quantification is now commonly performed using scanning microfluorometry (see Chapter 15). The basic technique has been extended to fluorescence induced with glyoxylic acid (Redgrave & Mitchell, 1982; incidentally they used a darkground illumination system, which is not recommended for quantification).

Serotonin (5-hydroxytryptamine, 5-HT) presents particular problems because its formaldehyde-induced fluorescence fades very rapidly, and the method does not appear to be as sensitive as one might wish (Tilders, Ploem & Smelik, 1974; Enerbäck & Jarlstedt, 1975; Geyer, Dawsey & Mandell, 1978; Jonsson et al., 1975). Geyer et al. (1978) in model experiments found that the amount of fading could be used as a measure of serotonin level, even in the presence of other fluorophores, and applied this technique to brain tissues.

Fluorescent probes

Some dyes are strongly affected by the environment of the dye molecules; measurements of changes in fluorescence parameters such as intensity, spectrum or polarization, give information about the region where the dye is situated. Such probes are particularly useful if they can be used in intact cells or organelles. Useful reviews of fluorescent probes include those of Haugland (1989), Watson (1987), Waggoner (1986), Stoltz & Donner (1985), and Beddard & West (1981). For suitable substances see Rost (1990) and Haugland (1989).

Fluorescent probes are relatively small molecules which show changes in one or more of their fluorescence properties as a result of interaction with their molecular environment. Such interaction may be related to adsorption onto or covalent binding to a protein or other macromolecule, or incorporation into a non-polar region of a membrane. The information potentially available from fluorometry of fluorescent probes includes probe environment, polar/non-polar nature of substrate, pH, redox potential, electric fields, concentrations, complexing, conformation change, distances, accessibility, rotations, lateral diffusion, group reactivities, and functional correlations. It must be noted that the substrate may be affected by prior fixation and by any toxic effect of the probe itself. Substances which are particularly useful as probes are the carbocyanine dyes (as probes of membrane potential), Fura-2 and other indicators of calcium concentration, anilinonaphthalene sulphonate (ANS) (as a membrane probe), and Acridine Orange (for studying nucleic acid conformation).

Fluorochromes currently used as quantitative probes for intracellular ions respond in two ways to increasing concentration of the relevant ion. The quantum efficiency and therefore the brightness of the fluorescence increase, and there is a substantial spectral shift in either the excitation or emission spectra. Measurements are therefore required at two wavelengths. The so-called 'dual excitation' fluorochromes, such as Fura-2 and BCECF, require measurements at two excitation wavelengths; whereas the 'dual emission' fluorochromes such as Indo-1, DCH and FCRYP-2, require measurements at two emission wavelengths. For such measurements, Kassotis et al. (1987) described a microfluorometer with provisions for rapid changing between two excitation wavelengths; the Joyce-Loebl 'MagiCal' apparatus allows very quick changing of excitation and/or emission filters (see Chapter 15).

Probes for membrane potential are electrically charged lipophilic molecules. For a review, see Waggoner (1985). The first such molecules to be recognized as probes for membrane potential were the *cyanins* (Sims et al., 1974). Cyanins are cationic, and partition in lipids according to the length of their side chains and the potential across the cell membrane. They partition strongly into negatively charged compartments such as the interiors of cells and mitochondria. Depolarization of the cell membrane of uncoupling of mitochondria allows the cyanine to leak out, with consequent loss of fluorescence. Substantial uptake of cyanins by mitochondria is a nuisance for measurement of cell membrane potential. *Rhodamine 123* is another cationic dye, partitioning into electronegative compartments, particularly mitochondria (Johnson *et al.*, 1982; Weiss & Chen, 1984). *Oxonols* are anionic dyes, otherwise functionally similar but opposite to the cyanins. For measurement of cell membrane potential, they have the advantage of avoiding mitochondrial fluorescence.

There are two types of pH probe: those whose fluorescence characteristics change with pH, and those which partition differently into cell compartments with different internal pH. Fluorochromes which have been used intracellularly include 9-aminoacridine, Azidofluorescein, BCECF, 6-carboxyfluorescein, various dicarbocyanines, various Oxonols, Rhodamine 123, and Umbelliferone.

Intracellular pH can be determined by microspectrofluorometry of spectral shifts of indicator fluorochromes, whose spectra are strongly affected by changes in pH (Visser, Jongling & Tanke, 1979; see also Heiple & Taylor, 1982). For example, excitation spectral shifts with pH occur in 6-carboxyfluorescein, from about 440 nm in acid to about 490 nm in alkali. Measurements of pH are based on the ratio of fluorescence emission intensity with excitation at 496 nm to the intensity of fluorescence emission with excitation at 452 nm. Emission spectral shifts occur with 2,3-dicyano-1,4-dihydroxybenzene (Valet *et al.*, 1981), from 450 nm to 483 nm as pH increases (Kurtz & Balaban, 1985). Both carboxyfluorescein (Thomas *et al.*, 1979) and dicyanodihydrobenzene can be loaded into cells as esters.

Probes for calcium are used to determine flux of intracellular calcium ions (Ca^{2+}). The study of intracellular calcium has expanded enormously recently (see Meldolesi, Volpe & Pozzan, 1988; Tsien, 1988; and issues of *Cell Calcium*). The probes are weakly fluorescent substances which become strongly fluorescent on binding Ca^{2+}. A list of those used to date (1989) is given in Table 5.2. During the current decade a range of highly fluorescent Ca^{2+} chelators has been synthesized by Tsien and his colleagues (Tsien, 1980, 1988; Grynkiewicz, Poenie & Tsien, 1985). Of these, Quin-2 was the earliest successful intracellular Ca^{2+} probe (Tsien, Pozzan & Rink, 1984) and is still widely used. However, it was largely superseded by Fura-2 and Indo-1; the most recent are Fluo-3 and Rhod-2, developed for long-wavelength excitation, e.g. with an argon laser. These substances are loaded into cells as their acetoxymethyl esters (I). Microspectrofluorometry of intracellular probes quantifies spectral shifts due to the environment of the fluorochrome.

I Acetomethyl ester

Table 5.2. **Fluorescent calcium (Ca^{2+}) indicators**

For details of individual substances, see Haugland (1989) and Rost (1991b).

Substance	Colour	Excitation
Chlortetracycline	Yellow	Blue
Fluo-3	Green	Blue
Fura-1	Blue	UV
Fura-2	Blue	UV
Fura-3	Blue	UV
Indo-1	Blue	UV
Morin	Greenish-white	UV
Quin-2	Blue	UV
Rhod-2	Red	Green
Stil-1	Blue	UV
Stil-2	Blue	UV

Probes for other ions have been described. Probes now exist for ions of sodium, potassium, magnesium, zinc, and chloride (see Haugland, 1989).

6

Principles of microspectrofluorometry

This chapter amplifies the general principles of microspectrofluorometry already mentioned in Chapter 1. Chapters 7–10 deal with practical details of instrumentation, techniques, standardization, and applications, respectively.

Microspectrofluorometry (MSFy) is concerned with the measurement of fluorescence excitation and emission spectra from specific regions of specimens. The instrument required (a microspectrofluorometer) differs from a microfluorometer in having two variable monochromators, one each in the excitation and emission optical pathways. Emission spectra are determined by measurements made at varying settings of the emission monochromator, the excitation remaining constant; whereas excitation spectra are determined by measurements made at varying excitation wavelengths while emission is measured in a constant wavelength band.

The practical object of microspectrofluorometry is usually the identification of a substance by its fluorescence characteristics; for example, the technique has been commonly applied to the identification of neurotransmitter amines rendered fluorescent by formaldehyde. On the other hand, spectral measurements on a substance of known fluorescence characteristics, used as a probe, give information about the local conditions within the specimen, as in the metachromasia of Acridine Orange staining of nucleic acids (Rigler, 1966). Applications are described further in Chapter 10.

Previous overall reviews of microspectrofluorometry include my own (Rost, 1974) and some incidental but valuable discussions (Nordén, 1953; Rigler, 1966). The majority of reviews are devoted to instrumentation, techniques and applications (e.g. Wreford & Smith, 1982; Rost & Pearse, 1974). General principles of spectrofluorometry are discussed by Slayter (1970).

Microspectrofluorometry is a powerful technique for the investigation of fluorescence in regions of microscopic size such as single cells. Microspectrofluorometry plays a role between that of quantitative fluorescence microscopy (which can reveal the presence, localization and amount of a fluorescent substance) and analytical chemistry, which can specifically identify substances in tissue extracts but which can give only vague localization, depending upon the precision of the dissection used to obtain the tissue for chemical analysis. Microspectrofluorometry cannot at present give quite as specific and reliable identification as biochemical analysis, but it can be used for analysis in respect of an accurately defined anatomical region, which can be as small as a single cell or part of a cell, even down to a single fluorescing granule (see, e.g., p. 125; Marques & Rost, 1973).

Compared to measurement of absorption spectra, microspectrofluorometry offers greater sensitivity and greater specificity – the latter because both excitation and emission wavelengths can be monochromated. However, the measurement of both excitation and emission spectra is subject to a number of sources of error, which will be discussed in Chapter 9.

Excitation and emission spectra

The characteristic spectra of fluorescence are the *excitation spectrum* and the *emission spectrum*. The excitation spectrum is a plot of excitation wavelength against the total intensity of the emitted fluorescence. The excitation spectrum is measured by exciting at different wavelengths and measuring at each excitation wavelength the intensity of the fluorescence emission. The emission spectrum is a plot of the relative intensity of the emitted radiation at each wavelength in the emission spectrum. These spectra may be considered as probability distributions: the excitation spectrum is the distribution of probabilities that a photon of given quantum energy will be absorbed and give rise to fluorescence, and the emission spectrum is the distribution of probabilities that a photon will be emitted with a given quantum energy.

The characteristic skewed shape of fluorescence excitation and emission curves (see Fig. 6.1) can be easily explained in the light of the facts given above. An excitation spectral curve has a peak at the photon energy, or wavelength, corresponding to the energy difference between the ground state of the fluorophore and some favoured vibrational level of its first excited state (Fig. 6.2). Incident photons with insufficient energy to raise the fluorophore to its first excited state will not be absorbed to produce fluorescence, and therefore the curve falls away rapidly below the peak. On the other hand it matters less if an incident photon has more energy than is necessary, since

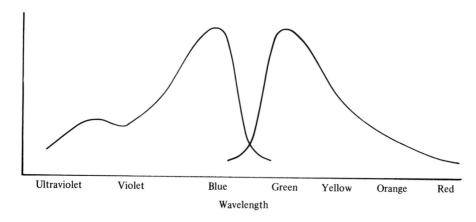

Fig. 6.1. Typical fluorescence excitation spectrum (left) and emission spectrum (right). Note the symmetry of the emission spectrum and the long-wavelength peak of the excitation spectrum; also the overlap of the spectra.

surplus energy can be converted into heat. Accordingly, the excitation spectrum has a long tail on the high-energy side, with additional peaks corresponding to second and higher excited states.

The shape of the emission spectrum is usually similar to that of the first absorption band (lowest peak of the excitation spectrum), but in 'mirror-image' form. Sometimes there are two peaks. The peak corresponds to the energy lost in the transition from the lowest vibrational level of the first excited singlet state (S_1) to some favoured vibrational level of the ground state (S_0; see Fig. 12.1). The curve falls off steeply on the high-energy side because of the improbability of the emission of photons with higher energy than that represented by the difference between the lowest vibrational levels of S_1 and S_0, since fluorescence is virtually always from the lowest vibrational level of S_1 and the energy of the molecule cannot fall below that of the lowest vibrational level of S_0. However, photons of much lower energy can be emitted, corresponding to transitions to higher vibrational levels of S_0. Accordingly, an emission spectrum has a long tail on the low energy (long-wavelength) side.

These considerations are of some importance in relation to the examination of mixed fluorophores. If two fluorophores (say A and B) are present, one (A) requiring excitation by a higher photon energy than the other (B), both can be excited by light of a sufficient quantum energy to excite A. If light of a low quantum energy is used, just sufficient to excite B, then A will be only very slightly excited (or not at all). Such circumstances occur regularly in tissue preparations: autofluorescence (proteins etc.) tends to require a high quantum energy for excitation and can therefore be reduced by using light at a quantum energy only just enough to excite the desired fluorophore. Increasing the quantum energy of the incident photons (decreasing the wavelength) beyond that necessary to excite the desired fluorophore is only deleterious, because of

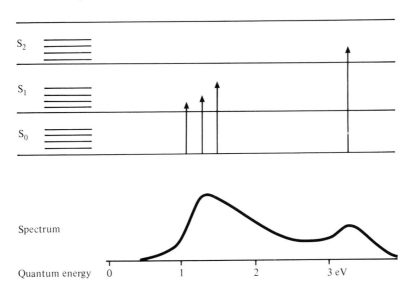

Fig. 6.2. Transitions between electronic states corresponding to various points of an absorption (or excitation) spectrum.

the increased autofluorescence and because the redundant energy can go only to heating the specimen.

The difference between the wavelengths of the excitation and emission maxima is known as the *Stokes shift*. This indicates the energy dissipated during the lifetime of the excited state before fluorescence. In principle it should be measured in energy units (eV), but in practice it is measured in wavenumber units, and may be calculated from the corrected maxima (expressed in nm) as:

$$\text{Stokes shift} = \left(\frac{1}{\lambda_{ex}} + \frac{1}{\lambda_{em}} \right) \times 10^7$$

Since fluorescence at room temperature occurs from the first excited state, the wavelength of emission (λ_{em}) should not be changed by altering the excitation wavelength (λ_{ex}). If, in practice, such a change is observed, it implies that a mixture of at least two fluorophores is present, each with different excitation and emission characteristics. As the excitation wavelength is changed, so the degree to which each fluorophore is excited varies, and so does the proportion which each contributes to the total emission. Such changes in the emission spectrum of a single substance in solution were until recently thought to indicate the presence of an impurity; however, Chen (1967a) reported that quinine and another substance in acid solution show shifts of their fluorescence emission-band spectrum with wavelength of excitation. Fletcher (1968) named these *B shifts*, and demonstrated them in several fluorescent compounds with rotatable auxochromic groups, in numerous solvents. He postulated that the B shift is due to the fluorescent molecule existing in at least two different 'average' conformations, each with its own distinct energy transitions. These conformations can be considered as being stabilized by the interaction of the solvent with the molecule in both its ground and excited states; the B shift then results from the excitation of different proportions of these two conformations.

The generally accepted relationship between the absorption and emission spectra, based on thermodynamic grounds, is the Kennard–Stepanov formula (Stepanov, 1957):

$$F(v) = bA(v) \, v^3 \exp\left(- hv/kT \right)$$

where $F(v)$ is the number of fluorescence photons per unit frequency (energy) interval, b a constant, v frequency, h the Planck constant, k the Boltzmann constant and T absolute temperature. At the time of writing (February 1990) the most recent practical application of this theory appears to be that of Björn & Björn (1986).

Instrumentation

I define a microspectrofluorometer as a fluorescence microscope equipped for the measurement of excitation and/or emission spectra from a defined region of the specimen. Microspectrofluorometers therefore differ from microfluorometers (described in Chapter 2) in having a light source of continuously variable wavelength, and/or means of measuring emission at continuously variable wavelengths (see Figs. 1.3 and 7.1). This usually implies the presence of at least one prism or grating monochromator; however, excitation spectra can be obtained with a tuneable laser, and this method is likely to be increasingly applied.

The instrumental requirements of microspectrofluorometry are essentially those of a microfluorometer (as described in Chapter 2) with the substitution of monochromators (giving variable wavelengths) for the excitation and emission filters, and a reference channel to monitor the output from the excitation monochromator at least during measurement of excitation spectra. The main components are therefore (1) a good microscope with epi-illumination, (2) a monochromatic light source of variable wavelength, (3) a reference channel to monitor the intensity of the light used for excitation, (4) a second monochromator system for analysing the emission from a defined area of the microscope field, and (5) a sensitive photometer. During measurements of excitation spectra, it is usual to employ a xenon arc lamp (which has a relatively continuous spectrum) and a conventional monochromator on the excitation side; on the emission side, the monochromating device may be a barrier filter, a broad- or narrow-band bandpass filter, or a monochromator. The purpose of the reference channel is twofold: to permit the calculation of corrected excitation spectra, and to compensate for fluctuations in the intensity of the lamp.

A variable-frequency (tuneable) laser is a modern alternative to the combination of a continuous-spectrum light source and a monochromator. It has the advantage of much greater spectral purity; also, in some systems, the wavelength can be changed more rapidly. A laser of fixed wavelength can be used for excitation during the measurement of emission spectra. Because excitation spectra commonly extend into the ultraviolet (UV), it is usually necessary to employ special UV-transmitting optics for the illumination system (see Chapter 3), including the coverslip over the specimen (assuming epi-illumination). Instrumentation is dealt with in more detail in Chapter 7.

Measurement of excitation spectra

The excitation spectrum is obtained by recording the photometer output for successive wavelength settings of the excitation monochromator at constant slit width, while maintaining the emission monochromator at a constant setting. The emission monochromator is normally set to the peak of the apparent (uncorrected) emission spectrum; alternatively, the emission monochromator can be replaced with a filter. The uncorrected spectrum so obtained must be corrected for the varying output of the excitation system at different wavelengths. In principle, assuming that the quantum efficiency of fluorescence is independent of excitation wavelength, the corrected excitation spectrum should be identical with the absorption spectrum of the same compound.

Excitation spectra should be reported in a standard manner (Chapman *et al.*, 1963). After correction, spectra should be presented as a plot of the relative number of photons emitted as a function of the quantum energy (or frequency, which amounts to the same thing) of the photons used for excitation. A scale linear in photon quantum energy or frequency is chosen to represent more accurately the relation between absorption and excitation spectra, and also to avoid compression of details at the short-wavelength end of the spectrum.

Measurement of emission spectra

The emission spectrum is obtained by recording the photometer output for successive wavelength settings of the emission monochromator at constant slit width, while maintaining the excitation monochromator at a constant setting. The excitation monochromator is normally set to the peak of the apparent (uncorrected) excitation spectrum; alternatively, the excitation monochromator can be replaced with a filter. The uncorrected spectrum so obtained is severely distorted, and must be corrected for several variable factors, namely: varying transmission of the emission system at different wavelengths, variable spectral bandwidth of the emission monochromator, and variable sensitivity of the photometer at different wavelengths. These are described in Chapter 9.

Emission spectra should be reported in a standard manner (Chapman *et al.*, 1963). After correction, spectra should be presented as a plot of the relative number of photons emitted as a function of quantum energy (or frequency, which amounts to the same thing). A scale linear in photon quantum energy or frequency is chosen to represent more clearly the relation between emission and excitation spectra.

The actual measurement of the emission spectrum can be made in several ways. One of the first ways utilized was that of photographing the spectrum produced by a prism; the negative was then measured with a microdensitometer. A better way is to pass the light through a monochromator to a photomultiplier tube; then either measurements of emission can be made at a number of discrete wavelengths and the results plotted as a graph, or the monochromator can be scanned through a range of wavelengths and the output from the photomultiplier plotted by a chart recorder. Other methods exist: e.g. the spectrum can be scanned by a video camera (West, Loeser & Schoenberger, 1960; Thaer, 1966b); or measurements can be made at a number of wavelengths simultaneously by a multichannel device such as a charge-coupled device (CCD). Experimental technique is discussed in more detail in Chapter 8.

Other applications

A microspectrofluorometer can be used for the same quantitative studies as a filter fluorometer, and is very versatile since the conditions of excitation and emission measurement can be adjusted to suit current requirements. However, the additional complication of a monochromator rather than a filter leads to some problems with homogeneity of field, so that for routine quantitative measurements a filter instrument is generally more suitable. Microspectrofluorometers can also appropriately be adapted for the measurement of fluorescence polarization and lifetime (see Chapter 12).

7

Instrumentation: microspectrofluorometers

For the purposes of this chapter, I define a microspectrofluorometer (MSF) as a fluorescence microscope equipped for the measurement of excitation and/or emission spectra from a defined region of the specimen. Microspectrofluorometers therefore differ from microfluorometers (described in Chapter 2) in having a light source of continuously variable wavelength, and/or means of measuring emission at continuously variable wavelengths. This usually implies the presence of at least one prism or grating monochromator: however, excitation spectra can be obtained with a tuneable laser (see Chapter 2), and this method is likely to be increasingly applied.

Principles involved in microspectrofluorometric instrumentation have been reviewed by Nordén (1953), Rost (1974), and Wreford & Smith (1982). Many of the basic principles of microfluorometers, as described in Chapter 2, also apply.

The optical arrangement of a microspectrofluorometer

The general arrangement of a microspectrofluorometer (Fig. 7.1) is basically similar to that of a microfluorometer, as described in Chapter 2, with the addition of variable

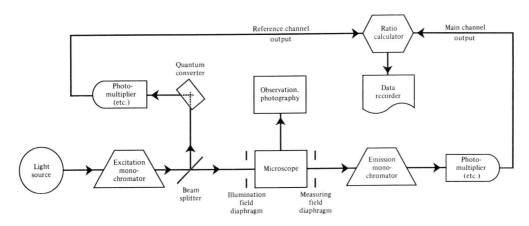

Fig. 7.1. Block diagram of a microspectrofluorometer. Thick lines show the light paths.

Table 7.1. **Field and aperture planes in a microspectrofluorometer**

This table lists various optical planes in a microspectrofluorometer in order from light source to photomultiplier, arranged in two columns according to whether the plane lies in a field plane (like the specimen) or an aperture plane.

Field plane	Aperture plane
	Light source
Field diaphragm and Köhler lens	
	M1 entrance and exit slits
	Condenser aperture diaphragm (condenser entrance pupil)
Specimen	
	Objective exit pupil
Ocular field diaphragm	
	Ocular exit pupil
M2 entrance and exit slits	
TV cathode	
	Photomultiplier cathode

Note:
M1, excitation monochromator; M2, emission monochromator; TV, television.

monochromating devices in the excitation and emission pathways, and a quantum converter in the reference channel to provide corrected excitation spectra. In setting up an instrument, it is useful to be aware of those optical planes which are conjugate with the specimen (field planes) and those which are conjugate with the condenser and objective apertures (aperture planes); these are listed in Table 7.1. It should be noted that the apertures of the optical components throughout each of the excitation, emission and reference pathways should be matched, so that the effective aperture of the system is not limited by a relatively unimportant component; for example the aperture of the lamp collector lens should be not less than those of the excitation monochromator and of the epi-illumination system.

The main components of a microspectrofluorometer may be summarized as follows; they are dealt with individually in detail below.

1. *A light source.* For measurement of excitation spectra, this must have variable wavelength over the required range, and is usually the combination of a xenon arc lamp and a monochromator, or may be a tuneable laser. During measurement of emission spectra, a constant wavelength is required, so the combination of a mercury arc lamp and a filter may be preferable to a xenon arc and monochromator, if one of the mercury lines is suitable for excitation.

2. *An illumination system.* This normally consists of a standard Köhler illumination system with epi-illumination. If a monochromator is used for selection of

excitation wavelengths, because of the long, narrow shape of the exit slit, an anamorphic system is required to convert this to fill the circular entrance pupil (aperture plane) of the illumination system (in practice, the 'exit' pupil of the objective, assuming epi-illumination). This appears to be best accomplished with a fibre-optic system using multiple fibres arranged in a column outside the monochromator slit and rearranged as a round bundle at the other end. The illumination system will incorporate a field diaphragm to limit the illuminated field, and a shutter to close off the illumination when light is not required.

3. *A subsidiary illumination system* using a substage condenser for normal transmitted light and/or phase-contrast observation. This is essential for locating the specimen prior to measurement. The system will include a shutter to close off the substage illumination when not required.

4. *A reference channel*, to compensate for variations in intensity of the light used for excitation, due to the spectral characteristics of the light source and of the excitation monochromator, as well as fluctuations in the intensity and position of the light source (see Chapter 9).

5. *A microscope stand* to support the specimen and optical systems.

6. *A magnifying system.* This includes the objectives and the observation ocular(s) and any projective used to project an image of the object into the measuring field diaphragm.

7. *A barrier filter and detector* for measuring the emission during measurement of excitation spectra. These normally consist of a conventional barrier filter followed by a photomultiplier and appropriate circuitry.

8. *For measurement of emission spectra*, the emitted light is first dispersed into a spectrum by a prism or grating. Thereafter there are three possibilities: (i) the spectra can be scanned past a slit behind which is a photomultiplier; or (ii) the spectrum can be scanned by a video camera; or (iii) the spectrum can be recorded as a whole by an array of photodiodes, or a charge-coupled device (CCD), or (in principle) photographically.

9. *A control and data logging system*, preferably based on a dedicated computer or microprocessor, to operate shutters in illumination and measurement pathways, change wavelengths of the monochromators, and to record data.

INSTRUMENTS

This section is concerned with current microspectrofluorometers; earlier instruments are described in Chapter 16. The following particularly good or useful features are to be noted in microspectrofluorometers described in the literature:

1. Continuously variable wavelength of the excitation light (e.g. Caspersson *et al.*, 1965) and interchangeable light sources (e.g. Björklund, Ehinger & Falck, 1968b).

2. Epi-illumination for excitation of fluorescence (Mellors & Silver, 1951; Böhm & Sprenger, 1968; Parker, 1969b; Rost & Pearse, 1971; Ploem *et al.*, 1974; David & Galbraith, 1975; Wreford & Schofield, 1975; Hirschberg *et al.*, 1979).

3. A reference channel (Olson, 1960; Caspersson, Lomakka & Rigler, 1965; Björk-lund, Ehinger & Falck, 1968b; Parker, 1969b; Pearse & Rost, 1969; Ploem *et al.*, 1974; Wreford & Schofield, 1975; Quaglia *et al.*, 1982).

4. Wavelength analogue output from a voltage divider attached to each monochro-mator (Thieme, 1966; Rost & Pearse, 1971; Wreford & Schofield, 1975).

5. Facility for rapid spectral scanning (Olson, 1960; West, Loeser & Schoenberg, 1960; Thaer, 1966b; Björklund *et al.*, 1968b; Pearse & Rost, 1969; Sprenger & Böhm, 1971a; Ploem *et al.*, 1974; Balaban *et al.*, 1986) or multichannel spectrometry for simultaneous detection of entire emission spectrum (Jotz, Gill & Davis, 1976).

6. Photometry by photon counting (Pearse & Rost, 1969; Rost & Pearse, 1971; Cova, Prenna & Mazzini, 1974; David & Galbraith, 1975).

7. Automatic digital data logging (Rost & Pearse, 1971; Cova *et al.*, 1974; Ploem *et al.*, 1974; Galbraith, Geyer & David, 1975).

8. Automatic immediate correction of spectra (Ploem *et al.*, 1974; Wreford & Schofield, 1975; Mazzini, Bottiroli & Prenna, 1975; Galbraith *et al.*, 1975; Klig, Demirjian & Pungaliya, 1976).

9. Light choppers for time resolution of luminescence (Parker, 1969b).

10. Laser excitation (Quaglia *et al.*, 1982; Ghetti *et al.*, 1985; see also Chapter 2).

11. Inverted microscope (Olson, 1960; Hirschberg *et al.*, 1979; Balaban *et al.*, 1986).

12. Provision for absorption measurement at the excitation wavelength (Olson, 1960; Rost & Pearse, 1971).

Current commercial equipment

Leitz

The MPV-3 and MPV-SP are the current Leitz photometric modules for use in conjunction with monochromators and a fluorescence microscope to form a microspectrofluorometer.

The Leitz MPV-3 (Fig. 7.2) is a modular system based on the Orthoplan microscope. Introduced in 1981, it superseded the MPV-2 and the microspectrograph-micro-spectrofluorometer.

Its main features are: (a) an epi-illumination system incorporating a monochroma-tor; (b) a photometric system incorporating a monochromator, a rotatable Abbé–König prism for image rotation, and a photomultiplier; (c) a normal microscope system of high quality; and (d) automatic computer control of monochromator wavelengths, stage scanning, fine scanning (by scanning mirror) within a field, image rotation, shutters in illumination and measurement pathway, and data acquisition and logging. The system is able to measure both excitation and emission spectra in the range 400–800 nm, and to record corrected spectra automatically.

The Leitz MPV-SP (Figs. 7.3, 7.4) is a similar modular system, based on the Orthoplan microscope, for measurement of emission spectra. It normally operates in the spectral range 400–800 nm (down to 220 nm with ultraviolet (UV) optics). The measuring modules incorporate a grating monochromator with a bandwidth of 1 nm and computer-controlled rapid scanning. Standard interference filters can be used instead (for excitation spectra). The electronic system provides automatic recording of corrected spectra.

Zeiss (Oberkochen)

The UEM microscope can be adapted for microspectrofluorometry. For measurement of emission spectra, a monochromator can be added to the photometric unit. For measurement of excitation spectra, a standard Zeiss monochromator can be

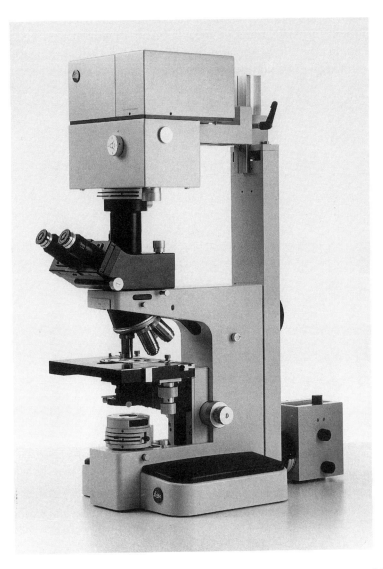

Fig. 7.2. The Leitz MPV-3 microspectrofluorometer. Photograph courtesy of Dr W.J. Patzelt, E. Leitz GmbH, Wetzlar.

incorporated into the excitation pathway; the detailed arrangement and the addition of a reference channel may be a matter for the individual. I have found the triangular optical bench system, which some Zeiss monochromators are designed to fit, convenient for the purpose; a carrier can be adapted to carry a lamphouse, and another to carry a projection lens.

MEASUREMENTS OF ADDITIONAL PARAMETERS

Measurement of absorption at the excitation wavelength can easily be arranged if dia-illumination is used; it is only necessary to tune the emission monochromator to the excitation wavelength, or to arrange for the emission monochromator to be bypassed via a filter to absorb fluorescence. A system of this kind was described by Olson (1960). Using epi-illumination, simultaneous measurement of absorption at the excitation wavelength is made possible by the addition of a third optical and measuring channel utilizing the substage illumination system in reverse: the condenser acting as an

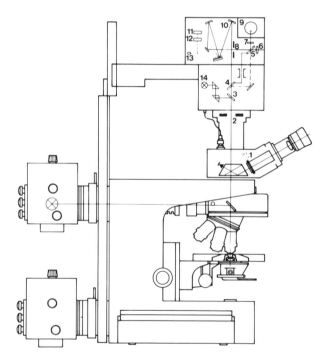

Fig. 7.3. Schematic diagram of a Leitz 'Orthoplan' microscope, with Ploem-Opak 2 epi-illuminator and MPV-SP photometer. The photometer system includes a grating monochromator for measurement of emission spectra. Key: 1, light flap; 2, entrance slit; 3, beam splitter for pilot lamp (14); 4, swing-in mirror for measurement without the monochromator (mirror 5 is swung out); 5, swing-in mirror for spectral measurement with the monochromator (mirror 4 is swung out); 6, magnifying lens for adjustment of slits (2 and 3); 7, filter turret; 8, exit slit; 9, photomultiplier; 10, monochromator; 11, incremental angle transducers; 12, drive for monochromator; 13, adjustment device; 14, pilot lamp. Diagram courtesy of Dr W.J. Patzelt, E. Leitz GmbH, Wetzlar.

objective (it can be replaced with an actual objective if required) collects light transmitted by the specimen from the epi-illumination system, and this can be measured by a photomultiplier (Rost & Pearse, 1971; Rost, 1973). A microspectrofluorometer can be adapted for the measurement of fluorescence polarization and decay times (see Chapter 12).

INDIVIDUAL COMPONENTS

Microscope

Since the instrument is, after all, primarily a fluorescence microscope, the microscope part of the instrument must be suitable for this purpose. Important requirements are: epi-illumination, with the possibility of simultaneous or alternative phase-contrast illumination by substage condenser; low light losses, particularly in the emission optical pathway; and provision for visual examination and photography of the measured field. Not all microscope stands are suitable, and the incorporation of a suboptimal microscope stand in a microspectrofluorometer is likely to be false economy.

Light sources

The basic requirements for the light source are high intensity, stability, and a continuously variable wavelength for determination of excitation spectra. An arc lamp

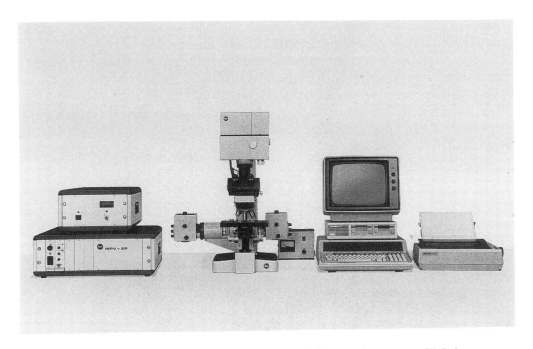

Fig. 7.4. Leitz MPV-SP on Orthoplan microscope. Photograph courtesty of E. Leitz GmbH, Wetzlar.

for use in conjunction with a monochromator is required to have a substantially continuous spectrum so that light of any wavelength can be selected; accordingly, xenon arc lamps are almost exclusively used in the measurement of excitation spectra. However a mercury arc lamp or laser may be more suitable for excitation during the measurement of emission spectra, a convenient line being used for excitation. A tuneable laser is ideal for the measurement of excitation spectra. For further details of lasers, see Chapter 2.

Monochromators

Monochromators are of two basic types: prism and grating (see Slayter, 1970). In a prism monochromator (Fig. 7.5), light from an entrance slit is first converted to a parallel beam of light by a collimator. This beam of light is dispersed by passage through a prism; because the refractive index of the prism varies with wavelength, light of different wavelengths is refracted at different angles, short-wavelength (e.g. blue) light being refracted more than long-wavelength (e.g. red). The phenomenon is identical with that which produces chromatic aberration in a lens. The dispersed beam is focussed by a focussing lens, and the desired wavelength band selected by a slit.

In a grating monochromator, the optical arrangement is similar, except that dispersion is produced by a diffraction grating. Such gratings are of two kinds: transmission and reflection. The grating can be cut in a special way ('blazed') so as to favour a particular band of wavelengths; for use with a xenon arc lamp Wreford & Smith (1982) recommend blazing at 300 nm to compensate for the reduced output of the lamp at short wavelengths.

Mirrors are sometimes used instead of lenses for collimation and focussing, to avoid chromatic aberration and difficulties with reduced transmission of prisms at UV and infrared wavelengths. Mirrors or reflecting surfaces in the prism are commonly used to arrange that the entrance and exit paths of the monochromator are coaxial. Further details of monochromators are given by Slayter (1970).

Graduated interference filters, whose spectral transmission varies along their length, are not as good as a prism or grating monochromator for emission measurement, and are not at all suitable for use on the excitation side.

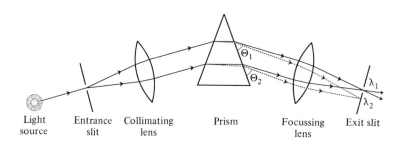

Fig. 7.5. Optical system of a prism monochromator. The exit slit can be replaced by a photographic film, video camera, or CCD.

Monochromator for excitation

This monochromator is required to produce a very pure monochromatic light for excitation: pure in the sense that the bandwidth can be quite wide by usual spectroscopic standards (e.g. 5 nm) but the proportion of light transmitted within the wavelength range of the fluorescence must be very low, of the order of 10^{-4} or less of the intensity at the excitation wavelength. Such spectral purity is difficult, if not impossible, to achieve with a 'single' monochromator (a 'double' monochromator consists of two monochromators connected to pass the same wavelength band; the light passes through both). The usual practice to date has been to employ a 'single' monochromator with supplementary glass filters. These glass filters can be very effective in reducing stray light, but suitable glass is not available for all wavelengths. The best glass at present seems to be UG1 (Schott) used for selection of the 365 nm mercury arc line. The ideal monochromator might be a double one, with a high aperture and wide slits (possibly some mechanism for removing the slits altogether when not required). The monochromator should be sealed to prevent ingress of dust particles; a few such fluorescing on an optical surface can spoil the spectral purity of the light.

The arc lamp is focussed into the entrance slit of the monochromator. The exit slit is then regarded as a virtual source for the illumination system. Because the slit is long and narrow, and the aperture of the illumination system is circular, there is some difficulty in matching the two. I believe that the best system is a bundle of optical fibres, arranged in a column at one end to match the slit and rearranged in a round bundle at the other end.

It is convenient if the monochromator gives an electrical analogue or digital output related to the wavelength, so that a data logging system (such as a computer) knows what excitation wavelength is in use. As an additional refinement, a stepping motor attached to the wavelength drive permits computer control of the wavelength.

The brightness of the excitation irradiation is maintained by choosing a high-aperture illumination system, and the most efficient monochromation system. For the latter, filters are usually preferable to a monochromator; if used, the monochromator should have a high aperture ratio. The density of the radiation passing a monochromator varies directly as the square of the slit width (see Chapter 9). However, the wider the slit, the less will be the spectral resolution. One has to compromise between sensitivity (with a wide slit) and specificity (with a narrow slit). A highly sensitive detection system will enable the slit width to be minimized, thereby increasing specificity as well as reducing the rate of fading.

The monochromator for emission generally has less stringent requirements than that used for excitation. Three basic designs are possible, in which the spectrum is scanned by (1) turning the prism or grating, or (2) turning a mirror to scan the spectrum, or (3) by viewing the spectrum directly with a video camera or equivalent (see Chapter 15). For measurement of emission spectra, a photomultiplier can be placed after the exit slit or the exit slit can be replaced with a photosensitive surface such as a photographic film, the cathode of a video camera, or a CCD.

A problem may arise in matching the shape of the measuring field to the monochromator entrance slit; as with the excitation monochromator, I believe that the best solution is a bundle of optic fibres. An Abbé–König prism can be used to twist the image of an elongate object to match the orientation of the monochromator slit, if it is not convenient to turn the specimen.

Reference channel

During the measurement of excitation spectra, the intensity of the light used for excitation changes as the wavelength is varied. These changes are due mainly to the spectral characteristics of the light source and of the monochromator. In order to compensate for variations in the intensity of the excitation due both to these spectral variations and also to instability of the light source, the intensity of the light used for excitation should be measured. This is most conveniently and appropriately arranged by placing a semi-transparent beamsplitter in the excitation pathway, to deflect a proportion of the light into a reference channel.

There are two ways of generating a reference beam: by a beamsplitter or by a mechanical chopper. A beamsplitter consists of a fixed sheet of quartz or a semi-silvered mirror placed in the illumination pathway at an angle of 45° so as to reflect a part of the light sideways into the reference system. The fixed beamsplitter method is simple and reliable, and permits simultaneous measurement in both channels. Allowance must be made for differing reflectivity of the beamsplitter at different wavelengths (see Chapter 9). In a mechanical chopper, a revolving shutter in the illumination pathway has alternately holes to allow light to pass to the object, and mirrors to reflect light to the reference channel; this was the arrangement in the Zeiss UMSP (see Chapter 16). The mechanical chopper method has two advantages: because the excitation is chopped at a steady frequency, a tuned amplifier can be used in the detector, with consequent improvement in the signal-to-noise ratio, and also the proportion of light reflected into the reference channel is substantially independent of wavelength. By a suitable arrangement of mirrors, the same detector can be used for both channels, which is an advantage for absorption studies but of no particular value for fluorescence. The main disadvantages of the chopper system are, firstly, that the reference and test channels are not measured simultaneously, and secondly that the chopper is only compatible with slow spectral scanning.

The reference channel should in so far as possible duplicate the illumination pathway, i.e. having an identical aperture, and an identical illumination field diaphragm. Placing the beamsplitter after the field diaphragm obviates the necessity for a matching field diaphragm and results in automatic compensation for any changes in its size. One way of correcting for varying reflectivity of a dichromatic mirror in the epi-illuminator would be to incorporate an identical one into the reference channel.

The reference beam is led to a quantum converter which absorbs the light at the excitation wavelengths and re-emits a constant proportion as light of a longer, constant wavelength. This is usually accomplished with a quartz cuvette containing a strong solution of a fluorescent dye such as Rhodamine B or Rhodamine (see Chapter 9). The linearity of the system is in part dependent upon the arrangement of the cuvette. The ideal arrangement seems to be that in which the incident light and the detector are at right angles, with the face of the cuvette at 45°, and with a mirror behind the cuvette or with a reflective coating at the rear of the cuvette.

Light emitted from the quantum converter is passed through a red barrier filter to a red-sensitive detector. It is convenient to have an iris diaphragm to vary the sensitivity of the detection system if required. Measurement of the light in the reference channel is generally most easily carried out with a photodiode, since the light is of long wavelength, at a high intensity level and does not normally vary as much in brightness as the emission-measuring channel.

Epi-illuminator

The beamsplitter may be either a '50/50' semi-transparent mirror, or a dichromatic mirror preferentially reflecting short wavelengths and transmitting longer ones. The latter has the advantage of being more efficient, the intensity of the measured emission being up to about three times greater than with a simple beamsplitter (see also Chapter 2). On the other hand, the dichromatic beamsplitters have complicated transmission curves, resulting in distortion of both excitation and emission spectra, and measurement of the required corrections is not easy. Therefore, for microspectrofluorometry it may be better to forego the advantages of a dichromatic mirror and use one with spectrally uniform transmission and reflection; a coating of stainless steel (Iconel) might be suitable. Metallic coatings for mirrors were discussed by Sidgwick (1979). If such a special coating is not available, I have found it best to use a simple quartz plate as a beamsplitter, as it has a known reflectivity and a high transmission (Rost & Pearse, 1971). Theoretically the light from the illumination pathway transmitted by the mirror could be used as a reference beam.

Objectives

To a considerable extent, the requirements for objectives are the same for microspectrofluorometry as for simple microfluorometry; however, generally a wider range of excitation wavelengths needs to be covered, extending further into the UV.

The essential requirements for the microscope objectives are: high transmission at relevant wavelengths (down to, say, 300 nm), very low autofluorescence, and preferably low glare (reflection of light backwards). These requirements are based on the assumption that epi-illumination is used; with a substage condenser, the objective is required only to have a high gathering power at the wavelengths of fluorescence.

Ordinary glass objectives may be quite suitable for use at wavelengths of 365 nm and longer, particularly those objectives designed with fluorescence epi-illumination in mind. For excitation at wavelengths below 360 nm, special objectives are required. I have found the Zeiss (Oberkochen) Ultrafluar series (Zeiss, 1967) satisfactory, although they suffer more from glare than do the normal glass objectives (which have coated air–glass surfaces). An alternative, practised in some laboratories, is to renounce epi-illumination at short wavelengths and use a quartz substage condenser. This also avoids difficulties with the epi-illuminator, but has other disadvantages (Rigler, 1966).

Detection system

Output from the emission monochromator is usually measured with a photomultiplier, which should normally be of red-sensitive type. Photomultipliers were described in Chapter 2. For rapid scanning of emission spectra, an oscillating mirror can be incorporated into the emission monochromator system, driven by a motor at appropriate speed. A television camera can be used instead (West *et al.*, 1960; Thaer, 1966b; Balaban *et al.*, 1986) or a CCD; either of these can be preceded by an image intensifier (Spring & Smith, 1987; see also Chapter 15). For a discussion of devices for measuring light at very low intensities, see Spring & Smith (1987) and Eccles, Sim & Tritton (1983).

A charge-coupled device (CCD) is a solid-state photometric device on a semiconductor chip. For details of these devices, see Chapter 15. A major advantage of the CCD for measuring emission spectra is that the entire spectrum is measured simultaneously. Their possible application to microfluorometry was discussed by Spring & Smith (1987).

Temperature control

Devices for cooling the specimen were described by Giordano *et al.* (1977) and Tiffe (1977). Temperature control at physiological temperatures is described in Chapter 13. Higher temperatures can easily be obtained with a heating stage.

Automation

A microspectrofluorometer is a complex device, and automation assists very much in obtaining accurate data. There should be an automatic process which closes a shutter in the substage illumination pathway, opens a shutter in the excitation pathway and simultaneously triggers a clock to record elapsed time from the commencement of excitation, and controls the scanning of an excitation or emission spectrum and the measurement and recording of measurements at each wavelength (Rost, 1973; Wasmund & Nickel, 1973; see also Wittig, Rohrer & Zetzsch, 1984).

Data-logging system

Microspectrofluorometry rapidly generates a substantial amount of data, which require processing (correction of spectra) before they become fully meaningful. Some sort of automatic data-logging system, and computer calculation of corrections, is required (Rost & Pearse, 1971; Rost, 1973; Cova *et al.*, 1974; Ploem *et al.*, 1974; Galbraith *et al.*, 1975; Wreford & Schofield, 1975; Klig *et al.*, 1976; see also Ritter *et al.*,

1981, and Wittig *et al.*, 1984). In my experience, the data-logging system should be capable of recording the general details of the experiment (as typed into a keyboard), and for each measurement, the wavelength, fluorescence intensity, and the time of the measurement (preferably elapsed time after commencement of irradiation).

FUTURE DEVELOPMENT

I imagine that the next generation of microspectrofluorometers will have:

1. A good-quality microscope with epi-illumination.
2. Excitation by a tuneable laser, the wavelength and output being under computer or microprocessor control.
3. A reference beam utilizing a fixed beamsplitter, a quantum converter, and a solid-state detector.
4. A scanning system to position the laser illumination on a chosen feature of the object.
5. Colour closed-circuit television for examination of the specimen and automatic video recording of the areas selected for spectral measurement.
6. An emission monochromator with its exit slit replaced by a CCD or similar device for detection.
7. Overall computer control of excitation, shutters, specimen scanning, measurement, correction of spectra and data logging and display. The spectra and the video image could be displayed on the same monitor, or on separate monitors.

8

Microspectrofluorometric technique

The process of measuring excitation and emission spectra consists of four stages: setting up the instrument; locating the specimen and defining the area from which measurement is to be made; setting the instrument for measuring the appropriate spectrum (excitation or emission), and determining the spectrum; and applying any necessary corrections to the apparent spectrum in order to obtain the corrected spectrum. This chapter deals with the process from the latter stages of setting up to completion of the measurement; earlier stages of setting up are dealt with in Chapter 3, and the application of corrections to spectra and data processing are dealt with in Chapter 9.

Setting up the instrument

The setting up of a microspectrofluorometer is very similar to the setting up of a filter microfluorometer, as described in Chapter 3, and is based on the setting up of the microscope as described in *Fluorescence microscopy* (Rost, 1991a). It is essential to check for homogeneity of illumination of the measuring field in terms of both intensity and wavelength; a similar criterion applies in respect of the measuring field. To some extent the measurement of the intensity of fluorescence of a small particle at various parts of the field tests both parameters; possibly the 'emission' spectrum of a reflective particle, such as a silver grain, could be measured at various parts of the field without a barrier filter to check on the spectral variations of the excitation light.

The monochromator, or both monochromators if two are used, often requires a supplementary glass filter to reduce light transmitted at wavelengths far removed from the nominal wavelength. In the case of the emission monochromator, during the measurement of emission spectra corrections must be applied for the transmission characteristics of this filter.

Specimen preparation

Because the excitation spectra are usually measured well into the ultraviolet (UV) range, it is usually necessary to use UV-transmitting materials for mounting the

Table 8.1. **Commercial mounting media**

Listed in ascending order of refractive index (n_D) at the sodium
D line (589 nm). Manufacturers' data except where marked
with superscript f.

Medium	Refractive index n_D
Water mounting medium[a]	1·455
Euparal[b]	1·481
Diatex[c]	1·490
Eukitt[d]	1·491
XAM neutral medium, improved white	1·491
Entellan	1·500
DePeX (Gurr)[a]	1·529
Canada balsam[e]	1·541
Caedax[e]	1·550
Neutral Mounting Medium (Gurr)[f]	1·485
Entellan new[f]	1·493
M.C.A. (Mount & Cover) (Gurr)[f]	1·493
Rhenohistol[e,f]	1·517
Canada balsam (Gurr)[f]	1·528

Notes:
[a] Hopkin & Williams.
[b] Farbenfabriken Bayer, Leverkusen.
[c] A. Svenssen, Malmö.
[d] E. Mertens, Bonn.
[e] E. Merck, Darmstadt.
[f] Data from Sebastian & Bock (1988).

specimen. Assuming epi-illumination, quartz coverslips are required; the slide can be of glass, which is very much cheaper, provided that it is non-fluorescent at all relevant wavelengths.

The mounting medium must also be UV-transmitting: water, glycerol, paraffin oil, and xylene are possibilities. A list of conventional mounting media is given in Table 8.1, and of possible fluids for temporary mounting in Table 8.2. A high viscosity discourages the coverglass from sliding about, and ideally the mounting medium should either dry to a solid mass or else be slow to evaporate. A buffer may be added to those fluids which are miscible with water. A reducing agent (anti-oxidant) may be added to reduce fading (see Chapter 11). Further details are given in *Fluorescence microscopy* (Rost, 1991b).

Locating the specimen

The same principles apply to locating the specimen for microspectrofluorometry as for microfluorometry (Chapter 3). To avoid fading, the desired region of the specimen should, if possible, be located without excitation of fluorescence; e.g. by conventional

Table 8.2. **Refractive indices of fluids available for immersion and as temporary mounting media**

In ascending order of refractive index (n_D) for the sodium D line (589 nm) or (n_E) for the iron E line (527 nm). All at 20 °C except where stated. Data from various sources. Refractive indices are given to only two decimal places; for more accuracy it appears necessary to measure the refractive index of the individual batch at the relevant temperature. See also Table 8.1, for commercial mounting media.

Fluid	Refractive index n_D
Air	1·00
Water	1·33
Propanone (acetone)	1·36
Ethanol	1·36
Ethanoic (acetic) acid	1·37
2-Propanol (isopropyl alcohol)	1·38
Olympus silicone immersion oil	1·40 at 25 °C
Paraffin Oil (B.P.)	1·44–1·48
Paraldehyde	1·40
Olive oil	1·47
Glycerol	1·47
Bergamot oil	1·47
1-Methyl-4-(1-methylethenyl)-cyclohexane (Histolene)[a]	1·47
p-Xylene	1·49
Methylbenzene (toluene)	1·49
m-Xylene	1·49
o-Xylene	1·50
Cedar-wood oil	1·52
Methyl benzoate	1·52
Ethyl iodide	1·52
Water	1·333
Olympus silicone immersion fluid	1·404 at 25 °C
Glycerol	1·47
Cargille[a] immersion oil, Type FF	1·4790 at 23 °C
Mersol immersion oil[b]	1·511
Cargille[a] Type DF	1·5150 at 23 °C
Cargille[a] Type A (low viscosity)	1·5150 at 23 °C
Cargille[a] Type B (high viscosity)	1·5150 at 23 °C
Cargille[b] Type NVH (very high viscosity)	1·5150 at 23 °C
Cargille[b] Type VH (very, very high viscosity)	1·5150 at 23 °C
Cargille[b] Type 37 (for use at 37 °C)	1·5150 at 37 °C
Zeiss (Oberkochen) immersion oil	1·515 at 23 °C ($n_E = 1·518$)
Olympus immersion oil	1·5155 at 25 °C
Lenzol[c] (cedarwood oil)	1·516
H&W immersion oil, tropical grade[a]	1·517
Microil immersion oil[c]	1·518
Uvinert[c]	1·519
Chlorobenzene	1·53
o-Cresol	1·54
Monobromobenzene	1·56
Benzyl benzoate	1·56

Table 8.2. *(cont.)*

Fluid	Refractive index n_D
Carbon bisulphide	1·63
Methylene iodide	1·74

Notes:
[a] Immersion oil: Fronine Pty Ltd, P.O. Box 380, Pennant Hills, NSW 2120, Australia.
[b] Immersion oil: Cargille recommend Type FF for fluorescence, having virtually no autofluorescence; Type DF recommended for highest resolution, meeting exact DIN specifications for refractive index and dispersion, but with slight ('very very low') greenish autofluorescence.
[c] Hopkin & Williams.

transmitted light microscopy, or by phase-contrast, interference-contrast, or dark-ground illumination, preferably using red light (which has only minimal effects on most specimens). If the required area can be found only by fluorescence, the light used for excitation should be reduced in intensity by means of a neutral-density filter. Under these latter circumstances it may be helpful to have two sources for fluorescence excitation: one giving illumination over the entire field, as in an ordinary fluorescence microscope; and the other for measurement only, incorporating a variable monochromator and with a limited field.

For the measurement, the illumination field diaphragm should be limited in size to avoid fading more of the specimen than necessary and to reduce scattered light. However, during the measurement of excitation spectra, the illumination diaphragm should always illuminate a somewhat larger area then the measuring field, to avoid errors due to variation in the size of the illuminated field, caused by chromatic errors in the illumination optics, during changes of wavelengths. For the measurement of emission spectra, for the same reason the measuring field should be larger than the illuminated field, i.e. the area from which the spectrum is measured should be determined by the illumination diaphragm, which is functioning at a constant wavelength.

Measurement of excitation spectra

The first problem is to decide the wavelength band in which to measure the emission, and how best to select that band. Selection of the wavelength range for emission measurement may be carried out in a number of different ways. In principle the emission may be monochromated by either a narrow-band device such as a monochromator or an interference filter or by a broad-band device such as a barrier filter. A narrow-band device can be centred on either the emission peak of the substance to be determined if known or on a wavelength somewhere in the long-wavelength tail of the emission spectrum. If a barrier filter is used it may be designed to pass either the major part of the emission or only a region of the long-wavelength tail. Combined together these possibilities give a total of four basic systems.

A narrow-band emission filter centred on or near the peak emission wavelength of the fluorophore gives a high degree of specificity. It also gives greater sensitivity than a filter of equal transmission and bandwidth and with peak transmission in the long-wavelength tail of the emission, where the intensity of emission is less. On the other hand the transmission curve of the filter is likely to overlap with the excitation spectrum, making it difficult or impossible to measure the longer wavelengths of the excitation spectrum.

A narrow-bandwidth filter with peak transmission somewhere in the long wave-length tail of the emission is the least sensitive of the four systems, having the double disadvantage of a narrow bandwidth and measurement in a wavelength region where the emission is less intense than at the peak. On the other hand, this system gives a somewhat higher degree of specificity than the use of a barrier filter with the same cutoff on the short-wavelength side. If an interference filter or a monochromator is used for this purpose it may suitably be combined with the barrier filter to give a sharper cutoff and therefore less overlap with the excitation.

A barrier filter transmitting almost all of the emission gives the greatest possible sensitivity, since the transmission of the filter is high and nearly all the emission spectrum is transmitted. However, even if the barrier filter itself has good rejection characteristics at short wavelengths there will still be a high likelihood of overlap of its transmission spectrum with the excitation spectrum.

A barrier filter with a cutoff wavelength somewhat longer than that of the emission peak may provide a good compromise between sensitivity and non-overlap with the excitation spectrum. It is moreover easy to set up, reproducible in the short term and requires a less complex optical system than does a monochromator. Also, its transmission is generally greater than that of a monochromator.

A monochromator can be used instead of the filters mentioned above. It is helpful to supplement this with a barrier filter, to provide sharper cutoff on the short-wavelength side. The easiest system to use is usually that of a monochromator with its slits wide open, functioning as a broad-band filter. With narrow slits, the monochromator functions as a narrow bandpass filter, and has the advantage of continuously variable wavelength.

On the excitation side, the illumination system is set up for illumination within the desired wavelength range for measurement of the excitation spectrum, with the monochromator slits set to the desired width and an appropriate supplementary filter chosen. The sensitivity of the reference channel is suitably adjusted to cope with the anticipated range of intensities over the wavelengths at which measurements are to be made; this is best checked by doing a dummy run with the light path to the specimen blocked to prevent fading.

When all is ready, the substage illumination is turned on to locate the specimen. After the desired region of the specimen has been placed in the centre of the field, the exact area to be measured must be placed within the measuring field. During the measurement of excitation spectra the field to be measured must be determined by a diaphragm in the emission pathway, to avoid any apparent changes due to chromatic error in the excitation optical system (imaging the excitation field diaphragm) as the spectrum is scanned. The field diaphragm in the illumination or excitation pathway must therefore be set slightly larger than the measuring diaphragm.

Before measurement commences, any light used for locating the specimen must be shut off, and ideally a light-trap should be placed above or under the substage condenser to minimize backward reflection of the excitation light which, having passed from the epi-illuminator through the specimen, passes into the condenser and may be reflected back up into the optical system again and add to the apparent background. This type of stray light is characterized by being greater in an empty field than with a non-fluorescent specimen present.

To commence measurement, a shutter in the excitation monochromator pathway should be opened and the excitation spectrum scanned. I have found it helpful to include a camera-type shutter in the excitation pathway, the flash synchronization contact (X-type for electronic flash) being used to initiate measurement. That is, unless the whole procedure, including the shutter, is under computer control. The spectrum either can be scanned continuously, the wavelength control of the monochromator being turned at a constant rate, or else a series of discrete measurements can be made at a range of preset wavelengths. The former method can be used to plot a spectrum directly on an *XY* recorder or a cathode-ray oscilloscope screen, and is particularly useful when an analogue output is obtained from the photomultiplier. If a digital output is obtained from the photomultiplier, it is necessary to make measurements at a series of discrete wavelengths, in which case it is best that the wavelength control of the excitation monochromator be advanced automatically by means of a stepping motor under computer control. In principle the light path could be closed by a shutter while the monochromator is being adjusted, in order to minimize fading.

The actual output from the photomultiplier needs of course to be corrected at each wavelength, by subtraction of dark current and by being divided by the reading of the reference channel, so that the excitation spectrum is plotted in terms of the relative quanta emitted per quantum of excitation (see Chapter 9). Because the emission pathway remains constant during the measurement, no corrections are required for the characteristics of the emission measuring device and associated optics.

Measurement of emission spectra

The procedure for measurement of emission spectra is similar to that for excitation spectra, except that this time the wavelengths at which emission occurs are scanned instead of the excitation wavelength, which remains constant. The detailed procedure given below is therefore very similar to, in places identical with, that for excitation spectra.

First, the wavelength band to be used for excitation must be decided. In principle the excitation may be monochromated by either a narrow-band device such as a monochromator or an interference filter, or by a broad-band device such as a combination of glass filters. A narrow-band device can be centred on either the excitation peak of the substance to be determined if known, or on a wavelength somewhere in the long-wavelength tail of the excitation. Because of the overlap between excitation and emission spectra, the excitation wavelength should normally be chosen at a relatively short wavelength, shorter than that corresponding to the excitation peak.

It is often convenient to use a strong spectral line from a mercury arc lamp for excitation; e.g. that at about 365 nm, which can be selected with Schott UG1 glass or an interference filter, either of which gives better transmission than a monochromator. Similarly the 405 nm and 439 nm lines can be selected with an interference filter, possibly supplemented with UG5 glass. The desired region of the specimen is optically isolated by a field diaphragm on the emission side, and light passed through a monochromator to a photomultiplier.

A narrow-bandwidth monochromator or filter centred on or near the peak excitation wavelength of the fluorophore gives the greatest possible degree of specificity. It also gives greater sensitivity than a monochromator or filter of equal transmission and bandwidth transmitting in the region of the short-wavelength tail of the excitation spectrum where the efficiency of excitation is less. On the other hand the transmission spectral range of the filter (centred on the peak of the excitation spectrum) is likely to overlap with the excitation spectrum, making it impossible to measure the shortest wavelengths of the emission spectrum.

If a monochromator is used, there is the problem that light leaving the exit slit has somehow to be focussed onto the region of the specimen. One method is to use the slit itself as the field diaphragm; an arrangement to limit the height of the slit may be required to enable the field to be adjusted to the dimension of the specimen. A disadvantage of this is that the field is not homogeneously illuminated, since there is a gradient of wavelength across the slit and therefore across the specimen; this, however, may not be of great importance for the measurement of emission spectra only.

A narrow-band filter or monochromator setting transmitting somewhere in the short-wavelength tail of the excitation spectrum gives the least sensitive of the four systems, having the double disadvantages of a narrow bandwidth and excitation in a wavelength region where the absorption is less than at the peak. However, it allows the full extent of the emission spectrum to be measured.

A broad-band filter providing excitation over most of the excitation peak of the fluorophore results in the greatest possible sensitivity, since the transmission of the filter is usually fairly high and strong excitation occurs. Moreover, the optical system required is simpler than that for a monochromator. However, even if the filter has good rejection characteristics at long wavelengths there will still be a high likelihood of overlap of its transmission spectrum with the emission spectrum, resulting in difficulty in measurement of the short-wavelength end of the emission spectrum.

The reference channel is suitably adjusted so that its sensitivity suits the intensity of the excitation; this is checked by opening the excitation light path to the reference channel beamsplitter, with the light path to the specimen blocked to prevent fading.

On the emission side, it is only necessary to ensure that the measurement system is set up for measurement within the desired wavelength range, with the monochromator slit set to an appropriate width and the appropriate supplementary filter chosen. The photometric system may need to be adjusted to an appropriate sensitivity setting.

When all is ready, the substage illumination is turned on to locate the specimen. After the desired region of the specimen has been placed in the centre of the field, the exact area to be measured must be placed within the measuring field. During the measurement of emission spectra the field to be measured should be determined by a diaphragm in the excitation pathway, to avoid any apparent changes due to chromatic error in the emission optical system (imaging the field diaphragm) as the spectrum is scanned. The field diaphragm in the measurement or emission pathway must therefore be set slightly larger than the measuring field. Care must be taken that the field diaphragm does not vignette the entrance slit of the monochromator. When all is ready, any light used for locating the specimen must be shut off, and ideally a light-trap should be placed above or under the substage condenser (as described above for excitation spectra).

To commence measurement, any light used to locate the specimen (phase-contrast, etc.) should be cut off, the shutter in the excitation pathway should be opened, and the emission spectrum scanned. The scanning can be initiated by a contact incorporated in the shutter mechanism. The spectrum can either be scanned continuously, the wavelength control of the monochromator being turned at a constant rate, or else a series of discrete measurements can be made at a range of preset wavelengths. The former method can be used to plot a spectrum directly on an XY recorder or a cathode-ray oscilloscope screen and is particularly useful if an analogue output is normally obtained from the photomultiplier. If a digital output is obtained from the photomultiplier, it is necessary to make measurements at a series of discrete wavelengths, in which case it is best that the wavelength control of the emission monochromator be advanced automatically by means of a stepping motor under computer control. In principle the excitation light path could be closed by a shutter in order to minimize fading while the monochromator is being adjusted, but I have rarely found this necessary.

The emission spectrum can alternatively be measured with a video camera or charge-coupled device; the latter has the advantage of measuring the entire emission spectrum simultaneously. A related system is the multichannel system of Jotz, Gill & Davis (1976). I shall not attempt to describe the practicalities involved in these systems, as I have had no experience with them.

General considerations

Background

There is invariably some background light, not forming part of the emission spectrum of the fluorophore. This may need to be determined and subtracted. Ideally, a non-fluorescent piece of tissue similar to the specimen should be used for this purpose; my experience is that this often actually gives a lower reading than a blank field, since it may block light reflected from below.

Fading (photobleaching)

Photobleaching affects the measurement of both emission and excitation spectra (see Chapter 11). Distortion of spectra occurs in two ways: fading during measurement of the spectrum will result in lower readings at the end of the spectrum measured last, and the spectrum may change as a result of photoreactions. The following summarizes procedures which may minimize the effects of photobleaching.

For purposes of comparison, measurements of fluorescence spectra should be made starting always at the same end of the spectrum, and proceeding in a standard manner; this method was used by Nordén (1953). Maintaining a constant direction of recording also minimizes complications due to gear backlash in monochromator drives. Comparisons are possible only between spectra on similar preparations measured with the same or identical instruments. For a given measurement time, it is advantageous to average spectra over a number of repeated fast scans rather than make a single slow scan (Cova *et al.*, 1986). If scanning can be achieved only slowly, another possibility is to measure the apparent spectrum repeatedly, noting the exact time of each measurement; the rate of fading at each wavelength is calculated and the results extrapolated backwards to estimate the spectrum at the commencement of irradiation. In my hands, this method has proved possible but time-consuming, and it is doubtful if the fading rate can be obtained with a high degree of accuracy.

The method of Ritzén (1967) in which the spectrum is measured after an initial period of irradiation appears to be justifiable if the fluorescent substance is pure, but does not give true results for a mixture. Ritzén used a standard irradiation: however, comparisons are possible only between spectra on similar preparations irradiated and measured the same way. The problem is that the spectrum may change (see Chapter 11).

It is evident that the most important factor is minimization of the irradiation. This requires that the detection system should be as efficient as possible: a high-aperture monochromator or high-transmission filters, a sensitive photomultiplier and a photon-counting electronic system seem indicated. With modern technology, measurement at each wavelength can be made with illumination by a short pulse from a laser, which may generate very little photodecomposition.

For the measurement of emission spectra, the use of a (second) monochromator is inefficient because measurements are made at only one wavelength at a time, emission at all other wavelengths being wasted. It may ultimately prove worth while to return to the photographic method (Rousseau, 1957; Ruch, 1960), using an image intensifier to enable the entire emission spectrum to be photographed with a reasonable exposure time (say 0·1 to 1·0 s). More likely, some form of multiplex spectrometry (Jotz *et al.*, 1976) may be appropriate, e.g. using a charge-coupled device. The literature to watch is that of the astronomers, who have the same problem of measuring spectra from faint objects. The problem of inefficiency does not arise with excitation spectra, since in measuring these the whole of the emission (in principle) is measured.

Temperature control

Temperature control at physiological temperature is described in Chapter 13. Higher temperatures can easily be obtained with a heating stage. Devices for cooling the specimen were described by Giordano *et al.* (1977) and Tiffe (1977), and their use described by Tiffe (1975) and Tiffe & Hundeshagen (1982). In my experience, one of the major difficulties is that moisture condenses onto the specimen and cold parts of the apparatus. A clear view of the specimen is facilitiated by the use of an immersion objective; at temperatures above freezing, a water-immersion objective is convenient as water condensation does not affect its use.

Data processing

Data processing and correction of spectra are described in Chapter 9.

9

Microspectrofluorometry: errors, standardization and data processing

Many aspects of errors, standardization and data processing in microfluorometry were discussed in Chapter 4. The present chapter deals with the additional problems which occur in microspectrofluorometry, associated with the measurement of spectra. Errors in microspectrofluorometry are of two basic kinds: errors due to the instrument not measuring equally at all wavelengths, and errors due to fading of the specimen during measurement. This chapter deals primarily with instrumental errors; those due to fading are dealt with in detail in Chapter 11.

Excitation and emission spectra as measured with any particular instrument are distorted by the optical and electronic characteristics of the particular instrument. Therefore, spectra as recorded do not correspond with the true spectra of the fluorescent substance, because of systematic and random errors introduced by the instrument. Excitation spectra are distorted mainly by variation in the intensity of light available for excitation at various wavelengths, while emission spectra are distorted mainly by the dispersion characteristics of the monochromator and by variations in sensitivity of the photomultiplier at different wavelengths.

It is therefore necessary to distinguish clearly between uncorrected (apparent) instrumental spectra on the one hand, and corrected spectra on the other. To enable comparison of results obtained in different laboratories, or even on the same instrument under different conditions, the results obtained must be corrected for all systematic errors, and the results plotted in a standard manner. The first section of this chapter deals with sources of error in (supposedly) corrected spectra. The method of calibration is dealt with in the second section of this chapter; the plotting of results was referred to in Chapter 6.

Errors in microspectrofluorometry

Instability of the light source
Instability of intensity and position of the light source was dealt with in Chapter 4. The only additional problem that might arise in microspectrofluorometry would be a change in spectral distribution, e.g. a drift in frequency of a laser used as a light source.

Non-monochromatic excitation

If the light source is a laser, the light available for excitation should be sufficiently monochromatic, provided that the generation of autofluorescence in the illumination system can be avoided. The combination of an arc lamp and a monochromator is more of a problem. Unfortunately, a 'monochromator' cannot be relied on to give truly monochromatic light or even light with a narrow bandwidth. There is inevitably some light of greatly different wavelength. This may be due in part to autofluorescence of dust particles on optical surfaces in the monochromator (Rost, 1972). Grating monochromators suffer from the additional disadvantage of transmitting also at multiples of the set wavelength; e.g. if a monochromator is set to give a first-order transmission at 400 nm, it will pass also at 800 nm and 1200 nm. These higher orders, and to a lesser extent autofluorescence in the optical system, can be blocked by supplementary filters. The filters normally take the form of coloured glass, which must be placed after the exit of the monochromator. A double monochromator (or two monochromators, one after the other) may be the best solution, except in respect of autofluorescence on optical surfaces.

Chromatic aberration in the illumination and observation optics

If the optical components of the illumination and observation systems are not fully corrected for chromatic aberrations, a change of wavelength will produce changes in the distances and magnifications of the images of the diaphragms and of the object. In consequence, during measurement of excitation spectra, the image of the illumination field diaphragm may change in position. Similarly, during measurements of emission spectra, the image of the object on the observation field diaphragm may change. To minimize these errors, the measuring field diaphragm should be on the illumination side for emission spectra and on the emission side for excitation spectra. Chromatic errors may be due not only to the type of error specifically named chromatic aberration (chromatic variation in lens power) but also to chromatic variations in field curvature.

Reflectivity and transmission of the epi-illuminator

Both the reflectivity and the transmission of the epi-illuminator mirror will vary with wavelength. Apparent excitation spectra are affected by spectral variation in reflectivity, since the latter affects the proportion of the illuminating beam which is reflected onto the specimen; emision spectra are affected by spectral variations in transmission. It is therefore necessary to calibrate the instrument for these characteristics. I have used a beamsplitter of plain quartz, giving a lower efficiency than a dichromatic mirror but with known reflectivity (described below under Errors in the reference channel).

Autofluorescence and foreign matter in the optical system

Autofluorescence in the optical system can be due either to foreign matter such as dust on optical surfaces, or to characteristics of glass and cementing elements in the system. Yellow and orange barrier filters, in my experience, are particularly liable to offend in this respect. Also, it is best to avoid imaging the surface of any field lens or mirror exactly into the object plane, since these surfaces may support particles of dust which will appear in the field.

Homogeneity of illumination

For homogeneous illumination of the object, the object plane must be illuminated by light which is of uniform intensity over the entire field, and if the light is less than perfectly monochromatic and therefore consists of a mixture of wavelengths, light of each wavelength must be uniformly distributed over the field.

The first of these requirements (uniform intensity) rules out the possibility of imaging the light source in the object plane (so-called 'critical' or source-focussed illumination). The second requirement (uniform distribution of all wavelengths present) rules out any similar arrangement for imaging the exit slit of a monochromator into the object plane. A Köhler-type illumination is required, together with a light source which is reasonably small, brilliant, and constant in intensity and position (Weber, 1965b). Assuming that a prism or grating monochromator is used in conjunction with a suitable light source, the obvious arrangement is to image the source into the entrance slit of the monochromator and to use the exit slit as a secondary source; a field lens is placed to image the exit slit into the aperture of the condenser (with epi-illumination, this is the 'exit' pupil of the objective), and the condenser is used to image a field diaphragm, near the field lens, into the object plane. Aberration in the condenser and field lens should be minimized.

The attainment of homogeneous illumination in the object plane depends mainly on (1) correct alignment of all optical components to the optical axis of the illumination system, (2) correct focussing of the collector lens (this varies with wavelength) and (3) correct focussing of the projection lens to image the exit slit of the monochromator into the aperture of the condenser (i.e. the 'exit' pupil of the objective, when epi-illumination is used). Tests for homogeneity of illumination are described in Chapter 3 and further discussed in Chapter 4.

Imperfect monochromation of the illumination

Unless a laser is used for excitation it is impossible to provide absolutely monochromatic illumination. For excitation during the measurement of emission spectra, the illumination need not be monochromatic so long as there is no significant amount of light within the wavelength range in which emission is being measured. For this purpose, I found that, in conjunction with a xenon arc lamp, the prism monochromator of the microspectrofluorometer of Rost & Pearse (1971) could be used with both slits fairly widely open, together with a piece of Schott UG1 glass (usually 1 mm thick) to filter off stray light in the visible region. On the other hand, for measurement of excitation spectra it is essential that the illumination should be essentially monochromatic: a bandwidth of about 0·02 eV is suitable if measurements are made at 0·05 eV intervals.

Errors in the reference channel

The function of a reference channel (Chapter 7) is essentially to compensate for variations in intensity of illumination caused either by instability of the light source or by changing the wavelength of excitation. Errors introduced by the reference channel therefore will be due to faulty optical arrangement, non-linearity in the quantum converter, or non-linearity in the reference channel measurement device and subsequent computational device. It is essential that the reference channel should receive a true sample of the light beam which illuminates the field; the optical system of the reference channel must have aperture and field diaphragms exactly analogous in size and position with those of the main illumination system.

The beamsplitter generating the reference channel should be placed in such a position as to sample exactly the beam of light passing to the object. The beamsplitter should be placed after the field diaphragm of the illumination system, and ideally after the aperture diaphragm also, although this latter arrangement seems possible only if an epi-illumination mirror is used as beamsplitter. For these reasons, in the microspectrofluorometer of Rost & Pearse (1971; see Chapter 16), the beamsplitter was placed after the illumination field diaphragm. Placing the beamsplitter before the diaphragm would have had the advantage that the reference channel reading would not have been affected by changes in size of the field diaphragm, but the reference beam would have included light not passing towards the diaphragm. It is important that the aperture of the reference channel should be the same as that of the main channel, for the same reason. Probably the ideal arrangement would be to use the epi-illumination beamsplitter to generate the reference channel (using for the latter, the light transmitted by the beamsplitter).

The beamsplitter can be relied on to reflect a strictly exact proportion of light into the reference channel, unless the light is so weak that stochastic variation in photon counts becomes significant (very unlikely). However, the proportion of light reflected will vary with wavelength, particularly if a dichromatic mirror is used. This is difficult to measure directly but can be calculated from the refractive index for the beamsplitter material. The relevant quantity is the ratio of the amount transmitted (to the main channel) to that reflected (into the beamsplitter).

I have used a quartz plate as a beamsplitter, relying on the light reflected from the two air–quartz interfaces for the reference channel. I laboriously calculated the amount which should theoretically be reflected at each wavelength, given the refractive index at various wavelengths (Table 9.1), and taking into account multiple reflections within the thickness of the beamsplitter (Fig. 9.1); however, on plotting the results, it appears that the amount reflected at 45° is directly proportional to the refractive index.

The quantum converter has received rather less attention than it should. As indicated in Chapter 6, the main requirements of the converter are that practically all the incident light must be absorbed and that the quantum efficiency of the solution must remain constant and independent of the wavelength of the exciting light. So long as these conditions are fulfilled, the emission from the cuvette as measured at a fixed wavelength band will be a relative measure of the intensity of the exciting light, irrespective of the wavelength of the latter.

The usual substance used as a quantum converter in microspectrofluorometry is Rhodamine B (C.I. 45170; tetraethylrhodamine) at a concentration of 3 g/l in ethylene

Table 9.1. **Reflectance of quartz beamsplitter**

Refractive index, and percentage reflected at various wavelengths. Incident and reflected light beams are assumed to be at 45° to the beamsplitter. Refractive index data by courtesy of Ernst Leitz GmbH, Wetzlar. Column 4 (R) data from Rost & Pearse (1971), calculated from refractive index; remaining columns calculated from R.

Wavelength		Refractive	% Reflected	% Transmitted	Ratio
nm	eV	index	R	T	T/R
313·17	3·959	1·4843	16·287	83·713	5·140
334·15	3·710	1·4798	16·103	83·897	5·210
366·32	3·384	1·4743	15·884	84·116	5·296
404·65	3·064	1·4696	15·692	84·308	5·373
435·83	2·845	1·4667	15·574	84·426	5·421
546·07	2·270	1·4601	15·305	84·695	5·534
587·56	2·110	1·4585	15·240	84·760	5·562
656·27	1·889	1·4564	15·155	84·845	5·598
768·00	1·614	1·4545	15·078	84·922	5·632

glycol (Ritzén, 1967). This has a high and constant quantum efficiency ($\phi = 0·89$) in the region 240–590 nm (Weber & Teale, 1957; Melhuish, 1962). Other quantum converters have been used, particularly for excitation spectra in the near infrared, e.g. for chlorophyll (Nothnagel, 1987; Duggan, DiCesare & Williams, 1983). The currently recommended substance, however, is Rhodamine 101 (rhodamine; Eaton, 1988).

In order to obtain almost complete absorption of the incident light in the Rhodamine B solution, both a high concentration and a sufficiently long light path are required. The most common size of quartz cuvette is 10 mm square; a larger cuvette, 15 mm × 10 mm, has been used to advantage, giving a longer light path in the 15 mm direction (Rost

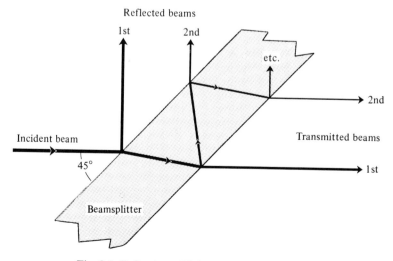

Fig. 9.1. Reflection of light by a beamsplitter plate.

& Pearse, 1971). This was found satisfactory up to about 500 nm. A reflective coating on rear faces of the cuvette increases the light path.

The usual solution of Rhodamine B, 3 g/l, gives significant reabsorption of emitted light. Errors due to this can be minimized by suitable placing of the cuvette. Ritzén (1967) placed the cuvette so that incident light fell on one face and measurements were made from an adjacent face; we (Pearse & Rost, 1969) initially followed this arrangement, but later (on the suggestion of Dr R. Rigler) arranged a cuvette at 45° so as to excite and measure from the same face of the cuvette. The reason for these arrangements is identical with the reason for using epi-illumination (see Chapter 1).

It is possible to check the operation of the quantum converter by comparing results obtained from a standard specimen with those obtained without the quantum converter and using a photomultiplier of known spectral sensitivity. A difficulty here is that the spectral sensitivity curve of the photomultiplier is sometimes expressed in terms of microwatts instead of quanta; in such case the curve must be converted to relative quanta by dividing each sensitivity value by the photon quantum energy at that wavelength (or by dividing by the frequency or wavenumber, or multiplying by the wavelength).

I suppose that one way of compensating automatically for varying reflectivity of a dichromatic beamsplitter in the illumination system would be to include an identical beamsplitter (if available) in the illumination pathway of the reference channel, but I have not tried this in practice.

The reference channel photometer should, in principle, contribute least to any errors in the system. The photomultiplier is operating under relatively good conditions: the light which it measures is of constant wavelength, because of the quantum converter system, so variations in sensitivity with wavelength do not affect the result.. The only problem is that the light emitted by the quantum converter is red, requiring a red-sensitive photomultiplier or solid-state device. The light emitted from the quantum converter should normally be very strong in comparison with that emitted by the specimen in the microscope, so that there should be no difficulty in achieving sufficient sensitivity and there should be negligible loss of accuracy due to photon statistics in a photon-counting system (see below). Virtually the only requirement for full accuracy in this part of the system therefore is a well-stabilized power supply and a stable amplifier system. Ideally, the reference channel photomultiplier should be powered by a separate high-tension power supply, independent of that used for the photomultiplier used for measurement of emission, so that changes in current drawn by either photomultiplier cannot affect the voltage applied to the other.

Background

Stray light at the emission wavelength derived from the illumination system (due to insufficient monochromation of the illumination) is difficult to deal with. The most important factor appears to be the design of the monochromator. Probably a double monochromator is necessary to reduce unwanted wavelengths to an acceptable level; the requirements for fluorometry are much more stringent in this respect than those of absorptiometry. Otherwise, it is necessary to supplement the monochromator with glass filters. The amount of this stray light which, with epi-illumination, is reflected back up the objective by air–glass interfaces can be minimized by selecting an objective which offends as little as possible in this respect. Some light from the epi-illumination may pass through the object and be reflected back by the condenser or from surfaces underneath the condenser: this stray light is greatest if the field is empty, and can be reduced by closing the substage condenser diaphragm and by placing a light-trap or black matt surface underneath.

Tissue autofluorescence presents another difficult problem. For quantitative fluoro-metry, its effect can be minimized by illumination at a wavelength corresponding to the excitation maximum of the specific fluorophore, which is thereby favoured in compari-son with other fluorescent substances present unless they happen to have similar excitation maxima. In analytical microspectrofluorometry this approach is often not possible, e.g. in measuring emission spectra the excitation wavelength usually has to be considerably shorter than that of the excitation maximum. Measurement of tissue autofluorescence from a control area was advocated by Ritzén (1967) but such an area is not always available. My own practice in most cases has been to use an empty field as the control area; this at least compensates adequately for stray light entering the system or reflecting from the objective (epi-illumination) but occasionally gives odd results, probably due to the fact that tissue reduces stray light reflected from the under-surface of the object slide (epi-illumination) or otherwise transmitted from below.

Transmission of monochromator

In measuring emission spectra, it is necessary to compensate for varying bandwidth and transmission at different wavelengths. For a monochromator of prism type, the relationship between the intensity of light transmitted and other parameters of the monochromator is as follows (Kortüm, 1962):

$$W = \frac{B \, S_\mathrm{w}^2 \, S_\mathrm{h} \, A}{f^2} \times \frac{\mathrm{d}l}{\mathrm{d}s}$$

where:

W = radiant energy per unit time at slit width S_w
B = radiant density at the entrance slit per wavelength unit
S_w = slit width
S_h = slit height
A = cross-section of the light bundle at the prism
f = focal length of the collimator
$\mathrm{d}l/\mathrm{d}s$ = linear (with respect to wavelength) dispersion of the prism

I believe that the same formula also applies if for 'wavelength' one reads 'frequency' or 'quantum energy'. In a prism monochromator the distribution is more linear with respect to frequency than with respect to wavelength, because of the dispersion characteristics of the glass or quartz in the prism, whereas a grating monochromator is linear in wavelength. It follows from the formula that the density of the radiation passing a monochromator varies directly as the square of the slit width.

Calibration of excitation and emission spectra

The calibration of microspectrofluorometers is important. Excitation spectra should be corrected absolutely, rather than relatively, to permit comparison with absorption spectra. It is less important for emission spectra to be corrected absolutely, so long as all instruments are corrected to the same standard specification.

The correction of fluorescence spectra has been discussed by Argauer & White (1964), Børessen & Parker (1966), Chen (1967b), Duggan *et al.* (1983), Lippert *et al.* (1959), Mayer & Thurston (1974), Melhuish (1962), Nothnagel (1987), Parker & Rees (1960), White, Ho & Weimer (1960), and Zalewski, Geist & Velapoldi (1982).

The correction of excitation spectra is carried out by measuring the intensity of the light used for excitation, by a standardized detector (Parker & Rees, 1960). A standard quantum converter in the reference channel has the desired effect and, as already indicated, is now fairly standard and permits comparison between laboratories. Correction is required for the varying reflectivity of the beamsplitter, i.e. the ratio of the amount transmitted (for the main channel) to that reflected (for the reference channel).

Absolute calibration of emission spectra is usually carried out by employing a standard tungsten-filament lamp of known colour temperature. Unfortunately this is technically somewhat difficult, requires expensive apparatus, and is not necessarily accurate particularly in the ultraviolet region where the emission of these lamps is low. Alternatively, any stable source of light can be used (e.g. a xenon arc) and itself be calibrated *in situ* by a detection system of known characteristics, e.g. a thermopile. See also Stair, Schneider & Jackson (1963).

By far the easiest way to calibrate both excitation and emission spectra would be to employ a standard fluorophore; spectra would be measured from the standard and compared with the known or assumed characteristics of the standard. Such a standard would be useful at least as a secondary standard. It might take the form of a solution of a pure fluorescent substance, held in a capillary tube after the manner of the fluorescence standards of Sernetz & Thaer (1970, 1973), or the fluorescent substance could be incorporated into a plastic block.

Calibration of excitation spectra

The design of reference channels has been described in Chapter 7. It is assumed that the quantum converter automatically corrects excitation spectra; and, even if it does not, at least the same type of quantum converter (using Rhodamine B) has been used by the major laboratories in this field and so results should be comparable between these. The only correction remaining to be made in connection with the reference channel is correction for the reflectivity of the beamsplitter at various wavelengths, as described above.

Calibration of emission spectra

Calibration of emission spectra is best carried out by measuring the apparent spectrum of a standard. For a primary standard, a standard lamp is used. The principle involved is as follows. The standard lamp is a tungsten lamp, the filament of which is heated by a constant current of known strength. Usually, the lamp is calibrated (e.g. by the National Physical Laboratory) in such a way that a specified constant current must be passed through the filament of the lamp to give a standard 'colour temperature'. This means that the temperature of the filament is such that its emission spectrum is the same as that of a perfect black body radiator at the specified temperature. The spectrum can be calculated or found from tables. An image of the standard lamp is formed at the position normally occupied by the specimen; the apparent emission spectrum is compared with the known true spectrum of the lamp, and appropriate correction factors calculated.

A standard fluorochrome solution has the advantages of reproducibility and low cost. The solution can be held either in a small open container, such as a well slide, or in a capillary tube. Microdroplets are more subject to fading. A solid material such as phosphor particles (West & Golden, 1976) or uranyl glass can be used.

An alternative approach, probably best used only as a check, is to determine the spectral characteristics of each item in the emission measurement system separately, and then combine the various corrections mathematically. This requires determining the spectral characteristics of the transmission of the optical system, the transmission of the epi-illumination mirror, the transmission of the barrier filter, the transmission of the monochromator (which will vary mainly with bandwidth), and the sensitivity of the photomultiplier (see Fig. 9.2). The technique is particularly useful, in my experience, if several barrier filters are in use at different times: their transmission characteristics can be held in a computer and appropriate corrections applied when required. Information concerning the transmission of the monochromator should be available from the manufacturers. The photomultiplier can be calibrated by the National Physical Laboratory; arrangements can be made for EMI photomultipliers to be calibrated by the NPL before delivery. In my experience, this calibration does not remain constant, and requires checking or re-calibration from time to time.

Fading during microspectrofluorometry

The measurement of excitation and emission spectra in the presence of fading (photobleaching) presents special problems, since more than one measurement has to be made from each area, unless measurements at all wavelengths are made simultaneously. If fading is rapid in proportion to the speed of scanning the spectrum, the measured spectrum will be distorted. If the specimen contains a mixture of fluorescent substances, these may fade at different rates, leading to changes in their combined fluorescence spectrum. Fading is discussed in detail in Chapter 11; practical procedures to minimize its effects are described in Chapter 8.

Models

Because of difficulties with absolute standardization, microfluorometry has come to rely mainly on the comparison of the fluorescence spectra of the object to be identified with that of a standard or model preparation, under stable conditions. The requirements for accurate models have been discussed by, among others, Ritzén (1967), Sernetz & Thaer (1970, 1973), and Jongsma, Hijmans & Ploem (1971); the matter is also dealt with, in a different context, in Chapter 4. To ensure proper comparability of standard and test objects, the following criteria (largely due to Ritzén, 1967; also Lichtensteiger, 1970, and Ewen & Rost, 1972) may be laid down in respect of the spectral standard:

1. Its size should be similar to that of the objects being measured, so that conditions of microscopy can be identical.
2. The standard must be in the same physical state as in the cell.
3. It must be possible to carry out the histochemical reaction on the model in the same way as on the tissue specimen.

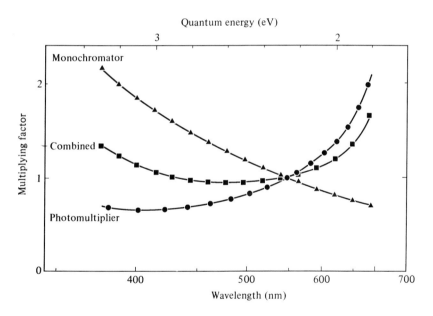

Fig. 9.2. Example of corrections applied to emission spectra, in the microspectrofluorometer of Rost & Pearse (1971). Triangles: corrections required for the varying dispersion of the prism monochromator. Circles: corrections required for varying sensitivity of the photomultiplier, as calibrated by the National Physical Laboratory. Squares: the combined correction for these factors. Lower scale: wavelength in nm. Upper scale: corresponding quantum energy (eV); this scale is linear, to straighten the monochromator curve.

Table 9.2. **Model systems for carboxyl-induced fluorescence of arylethylamines**

Carrier	Form	Author	Date
Gelatine, 5%	Layer	Corrodi & Hillarp	1963
BSA, 2%	Layer	Corrodi & Jonsson	1965
BSA, 1–10%	Droplets	Caspersson et al.	1966
BSA, 2%	Spots	Jonsson	1967
BSA, 0·1%	Spots	Rost & Polak	1969
Egg white albumin, 1–10%	Droplets	Caspersson et al.	1966
Egg albumin, 1%	Droplets	Van Orden	1970
Albumin, 2% in buffer	Spots	Björklund, Falck & Håkanson	1968
Albumin, 2% in buffer	Droplets	Björklund & Stenevi	1970
Albumin, 2% in buffer	Droplets	Björklund, Nobin & Stenevi	1971
Albumin, serum, 2% in buffer, pH 7	Droplets	Håkanson & Sundler	1971
Albumin, human, 2% in buffer, pH 7	Droplets	Lindvall & Björklund	1974
Albumin, human, 5%	Droplets	Agrup et al.	1977
Globulin, 1–10%	Droplets	Caspersson et al.	1966
Lysozyme, 1–10%	Droplets	Caspersson et al.	1966
Glycine, 0·02% in 2% sucrose	Droplets	Caspersson et al.	1966
Glycine, 0·02% in 2% sucrose	Droplets	Ritzén	1967
Glycine, 0·1% in 2% buffered sucrose	Spots	Björklund, Falck & Håkanson	1968
Glycine, 0·1% in 2% buffered sucrose	Droplets	Björklund & Stenevi	1970
Glycine, 0·1% in 2% buffered sucrose	Droplets	Björklund et al.	1971
PVP, 0·1%	Spots	Rost & Polak	1969
PVP, 0·1%	Spots	Cross, Ewen & Rost	1971
PVP, 1%	Spots	Rost & Ewen	1971
PVP, 0·1% in buffer, pH 6·4	Droplets	Ewen & Rost	1972

Note:
BSA, bovine serum albumin; PVP, polyvinylpyrrolidone.

4. If the concentration of fluorophore is high, the standard must be thin enough to ensure sufficiently uniform excitation throughout the entire layer, and to minimize the reabsorption of fluorescence emitted. This is particularly important if dia-illumination is used, but is less important if epi-illumination is used.

5. The standard is required to be constant, not fading during the measurement process and preferably also constant during storage.

6. The standard must be reproducible.

7. The carrier material must not show appreciable autofluorescence, which would mask specific fluorescence and distort apparent spectra.

8. Ideally, the standard should be easy to obtain or make.

Model preparations for spectral determination have been extensively used for microspectrofluorometric identification of arylethylamines (e.g. catecholamine neurotransmitters). In order to simulate conditions inside a cell, the substance of interest is incorporated into a carrier or matrix of dried protein or some substitute. Substances which have been used as matrices for models are summarized in Table 9.2. Ritzén (1967) used dried protein microdroplets as models for formaldehyde-induced fluorescence (FIF) of arylethylamines. The droplets were prepared by dissolving the arylethylamines in a protein solution, which was sprayed onto microscope slides and allowed to dry. The droplets were intended to mimic freeze-dried tissues as used for the histochemical reaction. Ritzén (1967) also made models with sucrose instead of the protein. Corrodi & Hillarp (1963), in their earliest experiments, used 5% aqueous gelatine; this substance, in my experience, is excessively autofluorescent and cannot be properly standardized. Bovine serum albumin (BSA) has been used widely; this material is a good medium for the Bischler–Napieralski reactions (e.g. FIF) but suffers, like gelatine, from excessive autofluorescence; Caspersson, Hillarp & Ritzén (1966) found emission peaks at 335 nm and at about 420–460 nm. Glycine in sucrose, pioneered by Caspersson *et al.* (1966), gives no detectable fluorescence and is chemically defined and therefore reproducible.

I have found that polyvinylpyrrolidone (PVP) is a suitable matrix for model studies (Rost & Polak, 1969; Rost & Ewen, 1971; Cross, Ewen & Rost, 1971). This substance (I) is chemically well-defined, has negligible autofluorescence and has no amino acid side-chains which might give unwanted reactions. It consists of a mixture of polymers with molecular weights in the range of approximately 10 000–700 000. It is a faintly yellow solid, resembling albumin, and soluble in water to give a colloidal solution. It is also soluble in ethanol.

$$\left[\begin{array}{c} \\ N \diagdown ^{O} \\ | \\ -CH-CH_2- \end{array} \right]_n$$

I

The concentration of the arylethylamine used in the models has typically been about 5 mM (in the original solution), and has varied from 50 μM (Rost & Polak, 1969; Cross *et al.*, 1971) up to 25 mM (Caspersson *et al.*, 1966). The final concentrations achieved in the dried droplets are not known.

For fluorochromes, it is also possible to use some of the standards which have been described in respect of microfluorometric mass determinations in cytochemistry, as described in Chapter 4. The best at present (for most purposes) is the capillary tube method of Sernetz & Thaer (1970, 1973); other standards include fluorochrome solutions in microdroplets or a cuvette, and fluorochromed beads of carbohydrate or latex. Unfortunately the fluorescence characteristics of a fluorochrome bound to a larger molecule (such as a protein) are not quite the same as those of the unbound fluorochrome, so that a solution of a pure fluorochrome is not an ideal standard for bound fluorochrome. A protein-bound fluorochrome standard was described by Goldman & Carver (1961).

Data processing

Correction of spectra is best carried out by a computer, preferably on-line to the photometer.

Correction of excitation spectra

The only corrections required for excitation spectra are:

1. Subtraction of dark current, if significant.
2. Division by reference channel level (after any required corrections for non-linearity of the reference channel).
3. Correction for reflectivity of the reference channel beamsplitter at different excitation wavelengths.
4. Correction for reflectivity of the epi-illumination beamsplitter at different excitation wavelengths; a different correction curve will apply for each beamsplitter in use, if more than one.
5. Subtraction of corresponding spectrum of background.

For each beamsplitter, and assuming that the beamsplitter reflectivity has been measured at a number of discrete wavelengths, the obvious way to obtain corrections is to store in the computer a table of wavelengths and corresponding reflectivities, and program the computer to interpolate between tabulated values as required. Unless the curve is very smooth, non-linear interpolation is required. However, speed of computation is much increased if the wavelength/reflectivity curve can be expressed as a polynomial. If a single polynomial cannot satisfactorily be fitted, two or three polynomials can be used, for different parts of the curve; with two, the program can use one below a specified wavelength and another at and above that wavelength. I think that these corrections could be eliminated by incorporating identical beamsplitters in the reference channel and epi-illuminator, giving automatic compensation.

Correction of emission spectra

The factors for which corrections are required are:

1. Dark current (unless negligible).
2. Reference channel value.
3. Transmission of emission optical system, including epi-illuminator beamsplitter.
4. Transmission and bandwidth of emission monochromator at each wavelength setting.
5. Transmission of any supplementary barrier filter used in conjunction with the emission monochromator.
6. Sensitivity of the photomultiplier at each wavelength.
7. Subtraction of corresponding spectrum of background.

Normally a single composite calibration curve is obtained to cover (3), (4) and (6). As with excitation spectra, the calibration curve is probably best expressed as a polynomial, otherwise non-linear interpolation is required.

Given digital data of spectra, if these are unimodal (i.e. show only one peak) analysis is possible by computing moments of the data. This methodology is conceptually easiest in respect of photon counting, where the moments of the photon quantum energies can be determined. The nth moment is the nth root of the mean of the nth powers of the observed counts at each photon energy. The first and second moments are respectively the mean and variance of the quantum energy (or frequency) of the photons. The third moment (skewness) indicates the asymmetry of the curve. The fourth moment (kurtosis) indicates the flatness of the curve. To be of any value, the data must be measured over the same range of wavelengths on all occasions. I attempted to use this technique for the photon-counting data of the instrument of Rost & Pearse (1971), but found that the moments (particularly the higher-order ones) tended to be unduly influenced by the shape of the tails of the spectra. Also, the curves are not those of a normal (Gaussian) distribution, but markedly skew, so some care may be required in interpretation. References to books on statistical analysis are given in Chapter 4.

Multivariate analysis is applicable to data in which multiple spectra have been measured from the same specimen. Such analysis usually relates to a series of emission spectra, measured either at different excitation wavelengths (Kohen, Hirschberg & Prince, 1989), or from successive regions of the specimen (Kohen & Kohen, 1977). The data can conveniently be handled as a matrix, in which data in each column is a measure of emission at a particular wavelength, and the rows represent emission spectra under different conditions. If each row corresponds to an excitation wavelength, the columns represent excitation spectra. The excitation/emission matrix can be used to estimate the number of fluorescent substances involved (Weber, 1961; Werner *et al.*, 1977; Kohen *et al.*, 1989).

10

Applications of microspectrofluorometry

In principle, microspectrofluorometry can be applied to any form of fluorescence: autofluorescence (fluorescence of untreated material), induced fluorescence (in which a substance in the specimen is converted to a fluorescent substance by a chemical reaction), or fluorochromy (where the tissue is stained with a fluorescent dye). A previous brief review is that of Rost & Pearse (1974). Like Chapter 5, the present chapter is intended to be illustrative rather than exhaustive.

In biology, microspectrofluorometry has so far been most widely applied to the identification of neurotransmitter amines and related substances, and to characterizing intracellular probes. Microspectrofluorometry has much to offer for forensic studies (Pabst, 1980), since it offers a non-destructive characterization of microscopically defined areas of specimens which can be quite tiny. In coal petrology, microspectro-fluorometry assists in quantifying the diagenesis of coals.

Autofluorescence

Autofluorescence is due to the presence of a fluorescent substance or substances in the specimen. In biological material, microspectrofluorometry offers the possibility of characterizing autofluorescent material; this should be particularly useful in species which have been relatively little studied. The technique offers possibilities in experimental pathology. My colleagues and I have studied the autofluorescence of amyloid (Pearse, Ewen & Polak, 1972) and of collagen and elastic fibres to distinguish them from nerve fibres (Rost & Van Noorden, unpublished). Spectra of elastin have been measured, with remarkably variable results, by various workers: Prenna & Sacchi (1964; Em 475 nm), Thornhill (1975; Ex 350 nm, Em 428 nm) and Deyl *et al.* (1980; Ex 340 nm, Em 410 nm); these appear to be all instrumental (uncorrected) spectra. Prenna & Sacchi (1964) also studied collagen (Em 450 nm). Banga & Bihardi-Varga (1974) studied atherosclerotic human aortae, identifying elastin (Ex 350 nm, Em 405 nm) and an 'autofluorescent component' (Ex 380 nm, Em 450 nm). Giordano, Prosperi & Bottiroli (1984) studied autofluorescence of muscle after injury.

Fluorescence of photoreceptor pigments is usually studied biochemically, but Ghetti *et al.* (1985) studied *in situ* the autofluorescence of a photoreceptor pigment in *Euglena gracilis.* Another recent study of autofluorescent pigments is that of Mello & Vidal (1985).

In petrology, microspectrofluorometry can be used to characterize components of coal in sections or polished slabs. With increasing diagenesis ('rank'), the emission maximum of liptinites usually shifts to longer wavelengths, i.e. from blue to red. Measurements of the emission spectra of coals, as sections or polished blocks, have been described by Homann (1972), Lo & Ting (1972), Jacob (1974), Ottenjann, Teichmüller & Wolf (1974, 1975), Stach *et al.* (1975), Stach (1982), Spackman, Davis & Mitchell (1976), Teichmüller & Ottenjann (1977), van Gijzel (1978), Crelling & Dutcher (1979, 1980), Crelling & Bensley (1980, 1984), Ottenjann (1980, 1981/82), Crelling, Dutcher & Lange (1982), Teichmüller & Durand (1983), Dobell, Cameron & Kalreuth (1984), Snowdon, Brooks & Goodarzi (1986) and Teerman, Crelling & Glass (1987). Microspectrofluorometry is suitable for determining the organic metamorphism and the expulsion stage of oil source rocks (Gutjahr, 1983). Identification of zircons by microspectrofluorometry is described by Aoki (1981, 1982a,b).

Recent and fossil pollen grains and spores show autofluorescence. The fluorescence spectra of microfossils have been studied by van Gijzel (1966, 1971, 1973, 1975, 1977, 1978). Variable fluorescence properties of foraminiferan shells may be due to selective uptake of trace elements from the environment during life (van Gijzel, 1966).

Parker (1969b) described the application of microspectrofluorometry to the characterization of oil droplets ingested by marine plankton following contamination of the sea by oil from ships.

Induced fluorescence

Induced fluorescence involves converting a non-fluorescent substance in the tissue into a fluorescent compound by treatment with a non-fluorescent reagent. Microspectrofluorometry has been widely used for the identification of arylethylamines ('biogenic amines') by spectral analysis of fluorescence induced with formaldehyde or glyoxylic acid. Identification of neurotransmitter amines requires measurement of excitation and emission spectra under carefully controlled conditions; suitable models containing the pure amine should be used as controls.

By microspectrofluorometry, in which the excitation and emission spectra from an area of aldehyde- or acid-induced fluorescence is determined, it is possible to assign the substance responsible to a particular group (catecholamine, tryptamine, *m*-tyramine). Some differentiation is also possible within groups, e.g. the differentiation of dopamine from noradrenaline by observing shifts in the excitation spectrum of formaldehyde-induced fluorescence after acidification (Björklund, Ehinger & Falck, 1968a,b; Wreford & Smith, 1979), or from the excitation spectrum of acetic-acid-induced fluorescence (Ewen & Rost, 1972).

The application of microspectrofluorometry to the differentiation of neurotransmitter amines was reviewed by Wreford & Smith (1982), Björklund & Falck (1973), Rost & Pearse (1973) and Ritzén (1967). Pioneering applications of this technique included studies of serotonin (Van Orden, Vugman & Giarman, 1965; Caspersson, Hillarp &

Ritzén, 1966; Hamberger, Ritzén & Wersall, 1966), catecholamines (Caspersson *et al.*, 1966; Hamberger *et al.*, 1966), *m*-tyramines (Jonsson & Ritzén, 1966; Jonsson & Sachs, 1971), and dopamine (Björklund *et al.*, 1968a,b; Ehinger *et al.*, 1968). Besides nerve cells, microspectrofluorometry of formaldehyde-induced fluorescence has been applied to serotonin in mast cells (Van Orden *et al.*, 1965) and to the formaldehyde-induced fluorescence of melanocytes and melanomas (Rost & Polak, 1969; Rost, Polak & Pearse, 1969, 1973; Cegrell, Falck & Rosengren, 1970). In embryological studies, the identification of serotonin and dopamine in the carotid bodies of chicks, quails and chick/quail mosaics (Pearse *et al.*, 1973) enabled tracing of developing neural crest cells in chick/quail chimaeras. Recent examples include the studies of Pohle, Ott & Müller-Welde (1984) and Scheuermann, Stilman & De Groodt-Lasseel (1988).

In pathology, microspectrofluorometry has been applied in the diagnosis of malignant melanomas, carcinoids, and phaeochromocytomas (see Fig. 10.1); the formaldehyde-induced fluorescence of these shows excitation/emission peaks at approximately 450/480, 400/540 and 410/480 nm, respectively.

The main problem in microspectrofluorometry of formaldehyde-induced fluorescence is fading (see Chapter 11), particularly pronounced in the formaldehyde-induced fluorescence of serotonin. Measurement of emission spectra alone is sufficient to differentiate the major groups of neurotransmitter amines; however, excitation spectra must be measured to identify the formaldehyde-induced fluorescence of melanomas. Excitation spectra including the ultraviolet (UV) range to about 300 nm or below are required to differentiate dopamine from noradrenaline.

Models are discussed in Chapter 9. For data on the excitation and emission spectra of formaldehyde-induced fluorescence from model experiments, see Björklund, Ehinger & Falck (1968b), Björklund & Falck (1973), Björklund, Falck & Lindvall (1975), Björklund, Nobin & Stenevi (1971), Corrodi & Jonsson (1966), Jonsson & Ritzén (1966) and Partanen (1978). For corresponding data on fluorophores induced

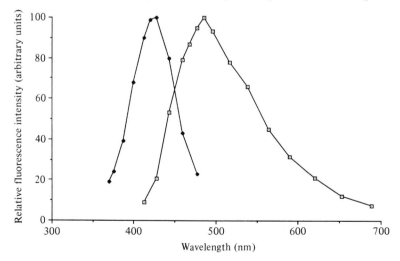

Fig. 10.1. Excitation (■) and emission (□) spectra of the FIF of cells in a phaeochromocytoma, confirming the presence of a catecholamine.

with gyloxylic acid, see Axelsson *et al.* (1973), Björklund *et al.* (1973, 1975) and Lindvall & Björklund (1974).

Microspectrofluorometry has also been applied to the characterization of fluorophores induced with *o*-phthalaldehyde (OPT).

Fluorochromes and probes

The most obvious application of microspectrofluorometry in the study of fluorochromes is the determination of excitation and emission spectra, to provide data for a logical choice of filters and other instrumental conditions for fluorescence microscopy (e.g. Bussolati, Rost & Pearse, 1969; Pearse, 1972). Instrumentation for this purpose can be quite simple, since precise determination of peaks is not required and excitation spectra need be measured only down to 350 nm.

Microspectrofluorometry of fluorochromes taken up by or injected into living cells can confirm the localization of the dye in particular regions, and if more than one dye is used, determine the proportion of each present. In an early application of microspectrofluorometry, Nordén (1953) confirmed the uptake of 3,4-benzpyrene by living cells, as part of a study on carcinogenesis; a recent example of a similar technique is that of Lautier *et al.* (1986). Identification and quantification of metabolites of introduced compounds is possible (Lahmy, Salmon & Viallet, 1984). Marques & Rost (1973) measured excitation and emission spectra from the intracellular fluorescent granules observed after UV irradiation of ascites tumour cells supravitally stained with nonfluorescent thiazine dyes, in order to investigate the origin of the observed fluorescence (see Chapter 11).

Fluorescent probes

As has already been said in Chapter 5, fluorescent probes are molecules which show changes in one or more of their fluorescence properties as a result of interaction with their molecular environment. Such interaction may be related to adsorption onto or covalent binding to a protein or other macromolecule, or incorporation into a nonpolar region of a membrane. Potentially, microspectrofluorometry of fluorescent probes can obtain information on such parameters as the conformation of the substrate molecules, the polar or non-polar nature of the substrate, pH and concentrations of other ions, redox potential, electric fields, complexing, conformation change, distances, accessibility, rotations, lateral diffusion, group reactivities, and functional correlations. Such probes are particularly useful if they can be used in intact cells or organelles. Reviews of fluorescent probes include those of Haugland (1989), Watson (1987), Waggoner (1986), Stoltz & Donner (1985), and Beddard & West (1981).

Metachromasia of fluorochromes such as Acridine Orange gives information about the substrate to which the dye is bound. Rigler (1966) studied metachromasia of Acridine Orange staining of nucleic acids, studying the conformation of the nucleic acids. I have employed microspectrofluorometry in the investigation of the fluorescence method for the demonstration of masked metachromasia using Coriphosphine O (Bussolati, Rost & Pearse, 1969; Maunder & Rost, 1972).

Fluorochromes currently used for measurements of the concentration of intracellular ions (such as H^+ and Ca^{2+}) respond in two ways to increasing concentration of the relevant ion. There is a substantial spectral shift in either the excitation or emission spectrum, as well as an increase in quantum efficiency and therefore in brightness. Development of such probes therefore requires microspectrofluometry, although in practical use ionic concentrations may be determined by microfluorometric measurements made at two wavelengths, as described in Chapters 5 and 15.

11

Photobleaching, photoactivation and quenching

This chapter deals mainly with the photochemical effects of the irradiating light on the specimen. Usually this leads to progressive diminution (fading) of fluorescence, but there are some reactions in which fluorescence is increased after irradiation. The terms photobleaching and fading are used roughly interchangeably to refer to progressive loss of fluorescence intensity during irradiation. Photobleaching is the more specific term, as fading can also be taken to refer to the effects of long-term storage. Photobleaching must be distinguished from decay, i.e. very rapid progressive reduction of intensity of fluorescence after cessation of irradiation, which is dealt with in Chapter 12. I therefore distinguish, in order of increasing lifetime, between *decay* after excitation, *photobleaching* during irradiation, and *fading* during long-term storage. *Quenching* is diminution of fluorescence, e.g. by chemical action on the fluorophore by another chemical species. *Recovery* is partial or complete restoration, after photobleaching, of the ability to fluorescence.

Photobleaching

It is characteristic of fluorescent preparations that they nearly always fade during irradiation. This is variously known as photobleaching, fading, photofading, and photodecay; the term photodecomposition is also used, implying (not necessarily correctly) destruction of the fluorescent molecule. Some preparations fade away into the background within minutes or less, while others are quite resistant. Photobleaching is usually a considerable nuisance, particularly in quantitative studies, in clinical diagnosis where two or more people may need to examine a preparation, and in photomicrography; on the other hand, photobleaching has been put to use in studies of molecular motion by examination of recovery after photobleaching. An excellent review of photobleaching was given by Picciolo & Kaplan (1984).

The mechanism of photobleaching is still largely unknown. I believe that several factors are involved, and that their relative influence varies from one specimen to the next and even within parts of the same specimen. Photobleaching is mainly due to photochemical reactions induced by the light used for excitation. The absorption of light, prerequisite for fluorescence, entails the raising of molecules to the excited state. The excited state is virtually a different chemical species, usually much more reactive than in

115

the corresponding ground-state molecule (see Porter, 1967). A small but significant proportion of the excited molecules, instead of fluorescing, undergo a photochemical reaction with the production of a new molecule which may be non-fluorescent, or at least non-absorbent at the excitation wavelength. It is commonly believed that photobleaching is due to a reaction between the excited fluorophore and oxygen to form a non-fluorescent product (Menter, Golden & West, 1978; Menter, Hurst & West, 1979; Giloh & Sedat, 1982; Vaughan & Weber, 1970). A non-oxidative mechanism is implied by the studies of Johnson *et al.* (1982), while Picciolo & Kaplan (1984) found evidence of both oxidative and non-oxidative mechanisms, on the basis of different photobleaching curves with and without an anti-oxidant and on the assumption that photobleaching in the presence of an anti-oxidant is unlikely to be oxidative.

Photobleaching may be due to a combination of photochemical destruction of the fluorophore, and changes in the quantum efficiency of the fluorophore. Simultaneous measurements of fluorescence emission and absorption, possible with the microfluorometer of Rost & Pearse (1971) showed a progressively greater diminution in fluorescence than in absorption, i.e. a loss of quantum efficiency (Rost, 1972). These results might be explained by the conversion of the fluorescent compound to a non-fluorescent but absorptive compound. A local heating effect might also be responsible for a loss of quantum efficiency, particularly initially until thermal balance occurs. At low temperatures (73·5 K) Tiffe & Hundeshagen (1982) found that photobleaching was due mainly to physical changes rather than to destruction of the fluorescent molecules.

The kinetics of photobleaching have been studied by several authors. The rate of photobleaching depends upon several factors, including the nature of the fluorophore, its chemical environment, and the intensity and quantum energy of the excitation. The initial kinetic characteristics, during one second or thereabouts, are different from those which are found subsequently. An example of a plot of fluorescence intensity over a period of time, showing photobleaching, is given in Fig. 11.1.

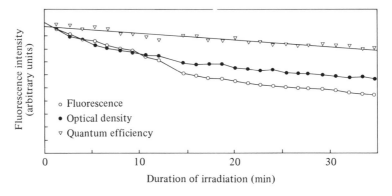

Fig. 11.1. Plot of fluorescence intensity against time, showing progressive reduction of fluorescence (fading). Open circles: fluorescence intensity. Solid circles: optical density (calculated from observed transmission). Triangles: quantum efficiency (calculated from observed fluorescence and transmission). Specimen: rat adrenal medulla, formaldehyde-induced fluorescence of noradrenaline. Results obtained with the modified Leitz microspectrograph of Rost & Pearse (1971; see Chapter 16).

The rate of photobleaching is characteristic of the fluorophore (Caspersson, Hillarp & Ritzén, 1966; Ritzén, 1967; Geyer, Dawsey & Mandell, 1978; Benson *et al.*, 1985). Some are particularly liable to rapid photobleaching (e.g. porphyrins) while others are relatively stable (such as ferulic acid in plant cell walls, uranyl glass, and the fluorochrome Acriflavine). In metachromatic staining, e.g. by Acridine Orange, the metachromatic (red) form fades more rapidly than the orthochromatic (green) form (West, 1965; Thaer, 1966b; Yamada *et al.*, 1966; Bussolati, Rost & Pearse, 1969). Differential photobleaching rates can be used to distinguish one fluorophore in the presence of another, by pre-bleaching labile fluorophores before measuring less labile ones (Hirschfeld, 1979), or by measuring the initial decay to quantify the labile fluorophore (Geyer *et al.*, 1978). If the specimen contains a mixture of fluorescent substances, these may fade at different rates, leading to changes in their combined fluorescence spectrum.

The rate of photobleaching is dependent on the intensity of irradiation, the rate constants increasing with increased energy density. This was first demonstrated visually with Vitamin A (Popper, 1944) and later with numerous other substances, and microfluorometrically with fluorescein isothiocyanate (FITC) (Goldman, 1960), Acridine Orange (Golden & West, 1974) and other dyes (Benson *et al.*, 1985).

Following studies of photobleaching over the period from about 1 second to 30 minutes of irradiation, photobleaching curves of various shapes have been reported in the literature (cf. Stoward, 1968b; Picciolo & Kaplan, 1984; Benson *et al.*, 1985). Benson *et al.* (1985) found photobleaching to follow first-order reaction kinetics; however, this does not appear to be always the case. Most commonly, photobleaching when examined over the period in question has been found to be rapid at first, fading more or less exponentially towards a constant minimum value; this is in accordance with my own experience with biological specimens. The rate constants in biological stained material were found to be spatially heterogeneous and varied within the same cell between 2- and 65-fold, depending on the fluorophore (Benson *et al.*, 1985).

Photobleaching has been found not to occur in the first few milliseconds of irradiation (Kaufman, Nesser & Wasserman, 1971; Enerbäck & Johansson, 1973) or even after up to 50 μs of irradiation (Bergquist, 1973). Fluorescence can therefore be measured or examined by excitation with a brief laser pulse without error due to photobleaching. Kinetics of photobleaching during the first 100 ms or so have been studied by Enerbäck & Johansson (1973) and Rundquist & Enerbäck (1976). The latter (Rundquist & Enerbäck, 1976) found an initial very rapid photobleaching, of about 10 ms duration, in three out of five fluorophores studied. These fluorophores were Berberine Sulphate on mast cell granules, Acriflavine-Feulgen stained mast cell nuclei, and green Acridine Orange staining of nuclei, but not the red Acridine Orange staining. This initial rapidly photobleaching fluorescence also showed rapid recovery.

The rate of photobleaching depends on the chemical circumstances of the fluorophore. Normally a bound fluorochrome fades more quickly than the unbound dye (Hirschfeld, 1979; see also Chapter 12). Photobleaching is diminished if the specimen is mounted in a medium containing an anti-oxidant (reducing agent) such as dithiothreitol, dithioerythritol, *p*-phenylene diamine, *n*-propyl gallate or β-mercaptoethanol (see below).

The extent of photobleaching is related directly to the amount of time during which fluorophore molecules remain excited, which depends upon the intensity and wavelength of the excitation, the optical density at the excitation wavelength, and the fluorescence decay time. The reasons for the non-exponential decay during photobleaching are far from clear. First, there may be two or more species present, e.g. bound fluorochrome and unbound fluorochrome. Normally the bound fluorochrome has a higher fluorescence efficiency than the unbound dye, a longer decay time, and therefore fades more quickly (Hirschfeld, 1979; see also Chapter 12). However, I believe (previously unpublished) that a possible reason for a non-exponential curve is as follows. If we assume that the quantum efficiency of the molecules is subject to stochastic variation, the molecules with a high quantum efficiency and corresponding longer decay time of the excited state will be likeliest to undergo photodecomposition; therefore, as photobleaching proceeds, the proportion of molecules with lower quantum efficiency will increase and the rate of photodecomposition will decrease. After the cessation of excitation, some recovery occurs if the specimen is kept in the dark, preferably in the cold. Recovery after photobleaching is discussed further below.

Photobleaching in fluorescence microscopy

The photobleaching of tissue autofluorescence appears to have been described first by von Querner (1932), who observed fluorescence in liver cells, fading under UV irradiation. This fluorescence was later ascribed to Vitamin A by Popper (1944), who described rapidly photobleaching green fluorescence of Vitamin A in tissue sections, and noted that the rate of photobleaching could be reduced by reducing the intensity of the irradiation. Cripps, Hawgood & Magnus (1966) reported on the rapid photobleaching of protoporphyrin. Falck & Owman (1965) observed a rapidly photobleaching red autofluorescence in the Harderian gland of the orbit in rat and mouse, presumably due to a porphyrin.

There are few published statements on the photobleaching of non-specific tissue autofluorescence. Prenna & Sacchi (1964) compared the photobleaching rates of elastic and collagen fibres, with and without formaldehyde fixation. Falck & Owman (1965), discussing the differentiation of formaldehyde-induced fluorescence of arylethylamines from autofluorescence, noted that autofluorescence was relatively stable to UV irradiation.

'Alteration' of autofluorescence in coal under irradiation is of some diagnostic value (Teichmüller & Ottenjann, 1977; Ottenjann, 1980, 1983).

Photobleaching of the formaldehyde-induced fluorescence (FIF) of catecholamines and tryptamines was noted by Falck (1962), who later observed that the FIF of catecholamines is less sensitive to UV irradiation than that of serotonin (Falck & Owman, 1965). This was confirmed by Caspersson et al. (1966), who found that the rate of photobleaching of the fluorescence was characteristic of the different groups of arylethylamines. They found that the FIF of serotonin faded twice as rapidly as that of catecholamines. In concentrations of up to 5% by dry weight of the protein medium, the photobleaching was practically independent of the amine concentration and of the reaction time with formaldehyde. Photobleaching curves were given for the FIF from noradrenaline (Ritzén, 1966a) and serotonin (Ritzén, 1966b). Differential rate of photobleaching was later used to quantify serotonin in the presence of catecholamines (Geyer et al., 1978). Rapid photobleaching of FIF from m-tyramine, metaraminol, α-methyl-m-tyrosine and m-tyrosine was noted by Jonsson & Ritzén (1966). Photobleaching of indolylethylamine FIF was studied by Jonsson & Sandler (1969); 6-hydroxytryptamine was found to give a more stable FIF than serotonin.

Stoward (1968a,b) studied the photobleaching rates of the fluorescence of mucosubstance salicylhydrazones and of their aluminium complexes. He found that the rate of photobleaching depended on the age and purity of the salicylhydrazide solutions used, and on whether the hydrazones had been treated with a weak or strong solution of aluminium salt. When the logarithm of fluorescence intensity was plotted against time, photobleaching appeared to be exponential in one or more stages, which were interpreted as representing photobleaching rates differing from one mucosubstance salicylhydrazone complex to another.

The photobleaching of fluorochromes used as supravital nuclear stains was studied by Eder (1986). In the case of the metachromatic fluorochromes Acridine Orange (AO) and Coriphosphine O, the metachromatic (red) form has been observed to fade more rapidly than the orthochromatic (green) form (West, 1965; Thaer, 1966b; Yamada et al., 1966; Bussolati et al., 1969). The photobleaching of Acridine Orange bound to glycosaminoglycans such as heparin has been studied in vitro by Menter, Golden & West (1978) and Menter, Hurst & West (1979).

The photobleaching of fluorochromes has been studied mainly in relation to quantification of Feulgen reactions by microfluorometry after staining with fluorescent Schiff and Schiff-type reagents. In Schiff-type reagents for Feulgen reactions, Prenna & Bianchi (1964) studied Acriflavine, Ruch (1966b) measured the photobleaching of Auramine O and bis-aminophenyl oxdiazole, and Böhm & Sprenger (1968) investigated Pararosaniline, Acridine Yellow, Acriflavine, and Coriphosphine O. Changes in the intensity and emission spectrum of Acriflavine–Schiff stained nuclei were studied by Sprenger & Böhm (1971b). Fujita & Fukuda (1974) found that, in specimens stained with the standard Feulgen procedure and irradiated with green light, the background staining faded more rapidly than the Feulgen staining; irradiation before microfluorometry therefore resulted in a greater difference between nuclear and background levels.

In relation to fluorometry of fluorescent antibodies, Goldman (1960) investigated the photobleaching of fluorescence of fluorescein in solution at various dye concentrations and light intensities. He found that photobleaching was more rapid with higher light intensities irrespective of concentration, and that more concentrated solutions faded more rapidly than the less concentrated ones. The slower the rate of photobleaching, the sooner was a level reached at which the fluorescence remained thereafter relatively constant during the remainder of the irradiation. Nairn et al. (1969) studied the effect of pH on FITC staining with UV and UV-blue excitation.

Photobleaching during microspectrofluorometry
The measurement of excitation and emission spectra in the presence of photobleaching presents special problems, since more than one measurement has to be made from each area. If photobleaching is rapid enough, in relation to the speed of scanning the spectrum, the measured spectrum will be distorted. If the specimen contains a mixture of fluorescent substances, these may fade at different rates, leading to changes in their combined fluorescence spectrum. Practical procedures to minimize the effects of photobleaching were described in Chapter 8.

Nordén (1953) and Ritzén (1967) both pointed out that photobleaching during recording of fluorescence spectra would cause distortion of the apparent spectrum. Nordén was sometimes unable to make intensity readings of benzpyrene fluorescence because of rapid photobleaching. To obtain consistent results, he always recorded emission spectra beginning with shorter wavelengths. Ritzén (1967) found that reasonably consistent spectra of FIF could be obtained if the specimen was irradiated for 3 min before starting the measurement.

Thaer (1966b), using a television system with the Leitz microspectrograph (see Chapter 16), demonstrated changes in the emission spectra of Acridine Orange-fluorochromed cells, within the interval from 5 to 45 s after the commencement of irradiation, the red component photobleaching more rapidly than the green. Photobleaching of the emission spectrum of serotonin FIF was investigated with a similar system by Vialli & Prenna (1969), who found that both emission intensity and the emission spectrum changed under irradiation. No explanation was offered for the changes in the emission spectrum, which should not occur with a pure substance. It is possible that the spectral changes were more apparent than real, being due to scanning a rapidly photobleaching spectrum of which the rate of photobleaching was also changing.

The problem of obtaining spectral excitation and emission curves from rapidly photobleaching FIF was discussed by Ritzén (1967), who stated that errors would occur if measurement took more than 5–10 s. Photobleaching of catecholamines was said to commence 5–10 s after commencement of irradiation, and to recover partially if the preparation was subsequently kept in the dark. He recommended making the measurement after 3 min of irradiation, as the rate of photobleaching was then less. Vialli & Prenna (1969) confirmed photobleaching of serotonin FIF and recovery in the dark, but did not observe any latent period before photobleaching commenced, and neither have I.

Cova *et al.* (1986) made a detailed study of the effect of scanning strategy on spectral measurement in the presence of photobleaching and other causes of intensity changes. They found that, to make best use of any given time available for measurement, it is better to average multiple scans rather than to make one slow scan.

For microspectrofluorometric identification of FIF in tissues, Ritzén (1967) proposed as one criterion of identification that the photobleaching rate of the specimen should be shown to be characteristic of the presumptive arylethylamine. A similar principle was used by Weissenböck *et al.* (1987) for the investigation of fluorescence in guard cells of plant leaves. For distinguishing between noradrenaline and dopamine, Wreford & Smith (1979) noted that after brief exposure to hydrochloric acid vapour, the noradrenaline-derived fluorophore showed a greatly increased rate of photobleaching. Photobleaching characteristics may be used to separate fluorophores, by prebleaching labile fluorophores before measuring less labile ones (Hirschfeld, 1979).

Methods for reducing photobleaching

Chemical means for reducing photobleaching are based on use of a suitable mounting medium. The use of an anti-oxidant in the mounting medium to reduce photobleaching has been recommended by several workers. p-Phenylenediamine, in either phosphate-buffered saline-glycerol or polyvinyl alcohol, retards the photodecomposition and increases the fluorescence of FITC (Gill, 1979; Johnson & Nogueira Araujo, 1981; Huff, Weston & Wanda, 1982; Johnson et al., 1982; Platt & Michael, 1983; Storz & Jelke, 1984; Valnes & Brandtzaeg, 1985) but was not found effective in reducing the photobleaching of TRIC (Giloh & Sedat, 1982). Sodium azide (NaN_3) and sodium iodide (NaI) have been found effective with FITC (Böck et al., 1985; Johnson et al., 1982). Picciolo & Kaplan (1984) recommended dithiothreitol (DTT) and dithioerythritol (DTE) for FITC. Other substances which have been found to reduce the photobleaching of FITC include sodium dithionate (Gill, 1979), n-propyl gallate (Giloh & Sedat, 1982; Valnes & Brandtzaeg, 1985), ascorbic acid (Giloh & Sedat, 1982), polyvinyl alcohol (PVA; Johnson et al., 1982), and 1,4-diazobicyclo(2,2,2)-octane (DABCO; Johnson et al., 1982); the usefulness of these latter was not confirmed by Böck et al. (1985). Remounting in PVA alone is recommended for long-term storage of sections mounted in PVA containing an antioxidant (Valnes & Brantzaeg, 1985). Another antioxidant which has been found effective is β-mercaptoethanol (Franklin & Filion, 1985) for nuclear staining with Acridine Orange, bisbenzimide, and Mithramycin. Photobleaching of the neuronal tracer Fast Blue is reduced by treatment with sodium nitroprusside (Spatz & Grabig, 1983) or (amounting to the same thing) the demonstration of a simultaneous horseradish peroxidase tracer by Mesulam's tetramethylbenzidine (TMB) method. Hamada & Fujita (1983) found that photobleaching could be reduced by incorporation of electron donors and molecules with SH groups into the medium.

Photobleaching of immunofluorescent preparations mounted in a buffered glycerol medium depends in part on the pH of the medium (Nairn et al., 1969). Mounting the specimen in a non-fluorescent resin reduces photobleaching and fading, probably by limiting the mobility of the excited molecule and giving it less opportunity to react with oxygen (Fukuda et al., 1980) (see Table 11.1).

Microscopic technique to reduce photobleaching of the specimen during examination can be applied in a number of ways. Focussing and preliminary examination of the specimen should if possible be carried out with light of lower quantum energy (longer wavelength) and lower intensity. Phase-contrast, interference contrast, and darkground illumination may be useful at this stage; red light should be used if possible (an additional advantage is that the observer's dark adaptation will not be spoilt by the red

Table 11.1. **Reagents reported to retard photobleaching and/or increase fluorescence**

Concentrations as used by Johnson *et al.* (1982).

Reagent	Concentration (g/l)	Remarks	References
p-Phenylenediamine	1		4, 5, 6, 7
Sodium azide (NaN$_3$)	25		1, 4
Sodium iodide (NaI)	25		1, 4
Polyvinylpyrrolidone (PVP)	25	Autofluorescent (1)	4
Polyvinyl alcohol (PVA)	25	Not confirmed (1)	4
1,4-Diazobicyclo(2,2,2)-octane (DABCO)	25	Not confirmed (1)	4
n-Propyl gallate	50	Not confirmed (1)	3
Sodium dithionate (Na$_2$S$_2$O$_4$)	3·5	Not confirmed (1)	2

Notes:

1, Böck *et al.* (1985); 2, Gill (1979); 3, Giloh & Sadat (1982); 4, Johnson *et al.* (1982); 5, Platt & Michael (1983); 6, Storz & Jelke (1984); 7, Valnes & Brandtzaeg (1985).

light). Photobleaching during examination can be reduced by selecting an appropriate wavelength and minimizing the intensity of the illumination. If an arc lamp is used for blue or green excitation, the filter should be supplemented with a barrier-type filter to eliminate shorter wavelengths. It may be possible to reduce the brightness of the excitation with a neutral density filter or by reducing the numerical aperture (NA) of the illumination system. If epi-illumination is not used, fading of the specimen can also be reduced by adjusting the field diaphragm so as not to exceed the diameter of the observation field. Techniques in microspectrofluorometry for minimizing the effects of photodecomposition during measurement of spectra are discussed in Chapter 8.

Photobleaching recovery

After photobleaching has taken place, some recovery may occur if the specimen is kept in the dark, preferably at low temperature (Ritzén, 1966a; Prenna, 1968; Dowson, 1984). At very low temperatures, recovery of fluorescence in crystalline material may be complete, that of biological material less so (Tiffe & Hundeshagen, 1982).

Photobleaching recovery and lateral diffusion

Fluorescence redistribution after photobleaching (FRAP; or fluorescence photobleaching recovery, FPR) is a technique for measuring two-dimensional lateral mobility of fluorescent particles (Peters *et al.*, 1974), for example the mobility of fluorescently labelled molecules on a cell surface. In principle, a defined region on the cell surface is irreversibly photobleached by exposure to an intense laser beam, and the subsequent recovery of the fluorescence is monitored microfluorometrically. To avoid damage to other components of the system, such as the nucleus, photobleaching is carried out at a wavelength which is not absorbed by those components and therefore does not affect them. Recovery of fluorescence is due to inwards migration of unbleached molecules, by diffusion or flow from the surrounding region. The kinetics of the recovery are related to the diffusion coefficient or flow velocity of the fluorescent molecules, and the degree to which the fluorescence regains the prebleach value gives the mobile fraction of the measured probe molecules.

Photobleaching is carried out either by a rapid flash, or continuously. In the former method, flash photolysis (Axelrod et al., 1976), the laser beam is initially heavily attenuated to a level which avoids significant photobleaching, then flashed at high intensity for a time much shorter than the characteristic time for motion of the relevant molecules across the illuminated region, and then returned to its low level to monitor the ensuing changes in fluorescence. This method gives results from a single flash and is easily interpreted. In the second method (continuous photolysis or photobleaching; Peters, Brünger & Schulten, 1981), the laser is kept at a constant, moderately high intensity. With this method, measurement takes place at higher light levels and is therefore easier, but interpretation is more difficult. Details of the latter method are given by Brünger, Peters, & Schulten (1985), and Scholz, Schulten & Peters (1985).

The photobleached region can be a spot, a line, or a more complex pattern such as a grid. Use of parallel lines for the photobleached field (Smith & McConnell, 1978) allows Fourier image analysis, giving particularly simple solutions of the diffusion equation and the possibility of detecting diffusional anisotropy (Smith, Clark & McConnell, 1979). Multipoint measurement within a bleached area enables immediate quantification of systematic flow and higher spatial resolution than the pattern method (Koppel, 1979). Excitation and photobleaching by a polarized light beam can reveal rotation of unbleached fluorophores into the plane of polarization (Smith et al., 1979, 1981). Apparatus was reviewed by Axelrod (1985) and described by several authors, notably Koppel (1979), who described in detail an instrument based on the Leitz Ortholux II microscope with epi-illumination, a laser, and a photometer with emission monochromator and photon counting; and Zs.-Nagy et al. (1984), who described an automated device based on an Olympus microscope. The mathematics involved were discussed by Koppel (1979), Smith & McConnell (1978), Axelrod et al. (1976), and Brünger, Peters, & Schulten (1985). Applications and techniques were reviewed by Koppel, Primakoff & Myles (1986), Jacobson et al. (1983) and Brünger, Peters, & Schulten (1985). Methodology is described in detail by Zs.-Nagy et al. (1984).

The technique has also been applied to the cell-to-cell transfer of fluorescent tracers (Wade, Trosko & Schindler, 1986; Safranyos et al., 1987). Recent examples of applications include those of Anders & Salopek (1989) and Zs.-Nagy, Ohta & Kitani (1989). The above description of fluorescence bleaching recovery is based on the review by Axelrod (1985) and summaries by Peters (1983), Jacobson et al. (1983), and Axelrod et al. (1976).

Fading during storage

The temporarily mounted preparations required for some studies, naturally, do not keep well. This category usually includes immunofluorescence, metachromatic staining (Acridine Orange and Coriphosphine O), and FIF. The last, in my experience, has the additional problem that background fluorescence increases on storage, even in material kept in a wax block. Preparations should be keep in the dark, preferably at a low temperature. Heimer & Taylor (1974) recommended storage of immunofluorescence preparations at -20 °C rather than 4 °C. More permanent preparations, mounted in a solidifying mounting medium (e.g. DPX) can be expected to keep at least for some months. Fading during storage is of course a problem with ordinary (diachrome) staining as well; see, for example, de la Torre & Salisbury (1962) on the fading of Feulgen staining.

Photoactivation

The excited state of a molecule tends to be much more reactive than its ground state, behaving as a quite different molecule. Some of the chemistry involved is discussed by Porter (1967) and Zweig (1973). Fluorescence photoactivation and dissipation (FPD; Ware *et al.*, 1986) is a technique related to FRAP.

Photoreactive dyes

While in general prolonged irradiation leads to fading of fluorescence, in some cases the excited form of the dye may undergo some reaction with the tissue, leading to firmer binding to the stained site. See, for example, the studies of Cox *et al.* (1984) on photoreactivity of phenanthridium dyes. The binding of photobiotin to other compounds is used as a means of labelling with fluorochromes.

Irradiation and toxicity

Excitation of molecules used as a vital stain can result in greatly increased toxicity. An application of this phenomenon is 'zap axotomy'; selective killing of all or part of single neurons achieved by filling the cell with a fluorochrome and irradiating the desired region with blue or UV light to excite the fluorophore (Miller & Selverston, 1979; Cohan, Hadley & Kater, 1983). The toxic effects of irradiation of fluorochromed nuclei in living cells was investigated by Eder (1986). *Escherichia coli* treated with Acridine and irradiated with ultraviolet show damage to both DNA and membranes, which may be lethal (Wagner, Feldman & Snipes, 1982). Parental injection of haematoporphyrin derivative (which is a mixture of porphyrins) followed by irradiation is used in the treatment of cancer. In excised corneas stained with a combination of Acridine Orange and Ethidium bromide, exposure to blue light excitation (10 mW/cm^2) or green light (25 mW/cm^2) for 2 min increased corneal swelling, and a decrease in endothelial cell density was noted subsequently (Kolb & Bourne, 1986). Inclusion of 10% calf serum in the staining and observation media had a protective effect.

Equipment for fluorescence microscopy can also be used for microirradiation of biological material, to investigate the effect on cells of localized irradiation with UV or visible light. For reviews of this topic, see Berns (1974), Berns & Salet (1972), Moreno, Lutz & Bessis (1969), and Zirkle (1957).

Irradiation-induced fluorescence

Prolonged irradiation with UV light has been observed to induce fluorescence not previously present, following vital staining with a non-fluorescent dye. In the course of an investigation into the uptake of cationic, non-fluorescent dye by tumour cells, Bastos *et al.* (1968) observed that supravital staining with a particular batch of Toluidine Blue resulted in a bright yellow fluorescence in cytoplasmic granules; after some minutes of UV irradiation, this fluorescence faded and gave way to a blue fluorescence (Fig. 11.2). A systematic trial with thiazine dyes showed that the yellow fluorescence could be observed only after staining with that one batch of Toluidine Blue, but that the blue fluorescence could be produced by prolonged UV irradiation of cells supravitally stained with any one of several thiazine dyes: Azure A, Azure I, Azure II, Methylene Blue, Thionin, Toluidine Blue (Bastos *et al.*, 1968; Marques *et al.*, 1968; Marques & Bastos, 1969). To this list was later added an oxazine dye, Brilliant Cresyl Blue (Marques & Rost, 1973). Excitation and emission spectra measured from single granules by Marques & Rost (1973) showed bimodal peaks, excitation at 340 nm and 394 nm and emission at 443 and 700 nm. We (Marques & Rost, 1973) concluded that the dye was taken up by the still living cells, and that with only limited oxygen supply in the preparation (between glass slide and a coverglass), the dye might be reduced to a leuco form. It was suggested that the UV excitation peak and the red fluorescence (Ex 340 nm, Em 700 nm) might be due to a metachromatic leuco dye, and that prolonged UV irradiation resulted in depolymerization to an orthochromatic form with blue fluorescence (excitation/emission peaks approximately Ex 395 nm, Em 443 nm).

Bisbenzimide (Hoechst 33258) is used as a fluorochrome for DNA, binding tightly to A-T regions. Incorporation of the base-analogue 5-bromodeoxyuridine (BrdU) into

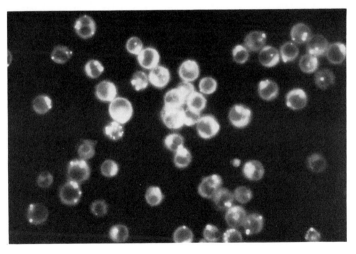

Fig. 11.2 Ultraviolet-induced fluorescence of Methylene Blue. Ehrlich ascites tumour of mouse. UV excitation, colourless barrier filter. Preparation by Dr D. Marques at the Royal Postgraduate Medical School, London.

newly synthesized DNA prevents fluorescence after bisbenzimide staining; however, this property is lost after a brief exposure to UV, so that the amount of DNA synthesized can be assayed by the increase in fluorescence after commencement of irradiation (Gordon & Parker, 1981).

Quenching

The term quenching is usually taken to include any phenomenon which leads to a reduction in the amount of fluorescence which might otherwise have been expected. Strictly, the term quenching is best used for the reduction of fluorescence by a competing deactivating process resulting from the presence of other molecules in the system. Four common types of quenching are recognized: temperature, oxygen, concentration and impurity quenching. The term quenching is also used to describe the alteration of the fluorophore by a chemical process, e.g. the borohydride reduction of FIF, and previous treatment which abolishes fluorescence for whatever reason. The quenching of fluorescence in solutions has been studied in some detail; see Guilbault (1973), Udenfriend (1962, 1969) and Pringsheim (1963).

Temperature quenching is a decrease in fluorescence as temperature increases. The degree of temperature dependence depends on the compound; it is usually about 1% per degree Celsius (Guilbault, 1973) but can be as high as 5% (e.g. tryptophan, Rhodamine B). The effect is believed to be due to increased molecular motion and increased frequency of collisions, resulting in increased probability of transition to the ground state before fluorescence can occur.

Impurities may decrease, increase, or have no effect on fluorescence. An excited molecule may transfer its surplus energy to an adjacent molecule, which itself becomes excited and may subsequently return to its ground state by any of the pathways which have been described above. If the absorbing molecule is normally fluorescent and the second molecular species non-fluorescent, the result is quenching of the fluorescence. Because transfer of energy from the donor to an acceptor molecule may be very efficient, and can apparently take place over distances as great as one micrometre, a very small concentration of an impurity can produce a high degree of quenching. On the other hand, if the donor molecule is non-fluorescent and the acceptor is fluorescent, the fluorescence of the acceptor may be greatly increased by the presence of the donor molecules if these are more efficient at absorption of photons. This latter process is known as *sensitized fluorescence*. Highly absorbing substances may interfere by robbing the fluorophore of light required for excitation.

A mixture of fluorophores will show a fluorescence emission spectrum which is the sum of the fluorescence emission spectra of the fluorophores present, in proportion to the intensity of fluorescence of each. In the absence of energy transfer, the latter will be determined by the excitation wavelength, the excitation spectra and quantum efficiencies of each fluorophore, and the relative concentration of each.

Oxygen quenching: one of the most notorious quenchers is molecular oxygen, which causes a reduction in fluorescence and completely abolishes phosphorescence. Oxygen in solution at a concentration of 1 mM typically reduces the fluorescence intensity by about 20% (Guilbault, 1973). In fluorescence microscopy, this probably accounts for some at least of the deterioration which usually occurs in fluorescent preparations on storage. It may be worth while including a reducing agent in mounting media. Oxygen quenching was studied by Kearns (1971), and discussed further by Wehry (1973). Quenching of fluorescence has been used for the measurement of intracellular oxygen concentration using pyrene-1-butryrate (Benson, Knopp & Longmuir, 1980; Podgorski *et al.*, 1981) or pyrene (Hargittai, Ginty & Lieberman, 1987).

Concentration quenching: at low concentrations of the fluorophore, as the concentration increases so does the intensity of fluorescence, because more light is absorbed. However, at very high concentrations, the fluorescence may be reduced, and the phenomenon is known as concentration quenching. Two mechanisms may be involved: absorption of light by superficial molecules may prevent deeper molecules from receiving the full strength of the exciting light; and if there is overlap of the excitation and emission spectra, some of the emitted light may be reabsorbed. Quantitative aspects of concentration quenching are dealt with in Chapter 4.

Prior treatment may be directed towards abolishing or altering the fluorescence in part of the specimen. An example is pre-staining of nuclei with Haematoxylin before staining with Thioflavine T for amyloid; the Haematoxylin prevents fluorescent staining of nuclei by the Thioflavine T, which (being a basic dye) would otherwise be the most prominent staining. Similarly Crystal Violet applied to isolated cells quenches surface but not internal immunofluorescence (Ma *et al.*, 1987). A more complex example is treatment of living cells with BrdU before staining DNA (as described above).

12

Time-resolved fluorescence, phosphorescence and polarization

This chapter deals briefly with techniques of fluorescence microscopy in which the fluorescence is characterized in terms of parameters other than excitation and emission wavelengths.

Time-resolved fluorescence

If fluorescence is excited by a sharp pulse of light, fluorescence emission takes place over a short period of time following excitation. Because the individual molecules emit at slightly different times, if the emitted light is measured with sufficiently fast apparatus, fluorescence can be shown to rise to a peak and then decay. The decay time of fluorescence (t) varies from one substance to another (Table 12.1) and this characteristic can be used in the identification of unknown fluorophores and also to permit the quantification of a fluorophore in the presence of a second fluorophore with similar or overlapping excitation and emission spectra but with a different decay time. Such measurements have become possible, and indeed routine, with the advent of pulsed lasers capable of providing pulses of picosecond (10^{-12} s) length (see Herrman & Wilhelmi, 1987) and photomultipliers with very fast response times. Time-resolved fluorometry is now widely used for chemical analysis. The technique is well explained by Wahl (1975). The application of this technique to fluorescence microscopy has the same advantage as the corresponding applications of fluorometry and spectrofluorometry, namely the possibility of characterizing a substance present in a defined region of a specimen (Brand & Gohlke, 1972).

Starting with a microfluorometer, measurement of fluorescence decay time in fluorescence microscopy requires the addition of standard instrumentation for the measurement of decay time. One such instrument, using an automatic pulsed laser, was described by Andreoni *et al.* (1980) and Docchio *et al.* (1984). Another instrument was described by Murray *et al.* (1986) and Morgan (1987); the instrument is commercially available (Philips Scientific & Industrial).

Hirschfeld (1979) pointed out the application of a principle, due to Einstein, that the fluorescent emission rate or its reciprocal τ_F are related to the integrated strength of the molecule's absorption band at which the fluorescence is excited. Unless this absorption band changes drastically in intensity, τ_F will be constant. In most fluorochrome stains, such drastic changes do not occur; absorption strength shifts far less than the fluorescence quantum efficiency (Udenfriend, 1969). It follows, reasoned Hirschfeld (1979), that significant quantum efficiency changes such as occur in fluorochrome stains must involve comparable changes in the overall lifetime (τ_E of the molecular excitation.

Consider the relation:

$$\phi = \tau_E / \tau_F$$

where ϕ is the quantum efficiency, τ_E is the lifetime of the molecule's excited state (reflecting not only its deactivation by fluorescence emission but also that due to any other radiationless process), and τ_F is the radiative lifetime of the fluorophore (the reciprocal of the excited molecule's emission rate). Quantum efficiency is determined by competition between fluorescent emission and radiationless deactivation back to the ground state.

Published estimates of fluorescence lifetime (τ) of substances vary, sometimes quite considerably. One cause of variation is that multi-component exponential decays may produce different apparent decay time, according to the method of data analysis. Therefore, as in all microfluorometry, a standard (Chen, 1974) should be measured along with the experimental material. Fluorescent lifetimes of fluorochromes attached to protein have been published by Chen & Scott (1985); see also Table 12.1. Normally the bound fluorochrome has a higher fluorescence efficiency than the unbound dye, a longer decay time, and therefore fades more quickly (Hirschfeld, 1979).

Applications

Time-resolved fluorescence microscopy has been applied to the study of chromatin, using as a probe Quinacrine (Arnt-Jovin *et al.*, 1979) and Quinacrine mustard (Bottiroli *et al.*, 1979, 1984; Andreoni *et al.*, 1979, 1980; Docchio *et al.*, 1986). The fluorescence decay of Quinacrine bound to a variety of DNA and chromosome preparations was studied by Arnt-Jovin *et al.* (1979). Analysis of the decay curves revealed at least two, probably three, components for the free dye at neutral pH, with τ approximately 1, 3 and 8 ns. Bonding to DNA resulted in increased τ, typically 1, 8 and 26 ns. They used these data to conclude that brilliantly fluorescent regions of Quinacrine-stained chromosomes may consist of regions with a high A-T content. Uptake of haematoporphyrin derivative (HPD) has been studied by Docchio *et al.* (1982, 1986) and Schneckenburger *et al.* (1985, 1987) using time-resolution of components.

Phosphorescence

Phosphorescence is similar to fluorescence, but occurs as the result of transitions (see Fig. 12.1) from the lowest triplet excited state (T_1), instead of the lowest singlet excited state (S_1), to the ground state (S_0). This process is less probable and therefore takes longer; hence, phosphorescence takes place over a longer period after excitation, long enough to be detectable by eye. Also, phosphorescence in general appears much weaker

Table 12.1. **Fluorescence lifetimes**

This table lists published values of the fluorescence lifetime (τ) for various substances in solution, showing visible fluorescence. The lifetime bound to tissue components may be different (e.g. see Arndt-Jovin et al., 1979; some values are given by Chen & Scott, 1985).

Compound	Solvent	τ (ns)	Reference[a]
Acridine Orange	Water	2·0	1
9-Aminoacridine	Ethanol (100%)	15·2 ± 0·2	2
1-Anilinonaphthalene-8-sulphonic acid	Water	2·3	1
Eosin Y	Water, 0·1 M NaOH	1·7	1
Fluorescein, 1 nM	Water, 0·1 M NaOH	4·5	1
Fluorescein	0·1 M NaOH	4·6 ± 0·1	2
4-Methylumbelliferone	Water	5·6	1
DPNH	Water, 0·1 M NaHCO$_3$	4·5	
TPNH	Water, 0·1 M NaHCO$_3$	4·3	1
Proflavin	Water	4·5	1
Protoporphyrin I	Water, 10 mM HCl	7·2	1
Pyronin G		4·5 ± 1·0	3
		2·9 ± 0·3	4
Quinacrine	Water	1, 3, 8	5
Quinine	Water, 50 mM H$_2$SO$_4$	19·0	1
Quinine bisulphate	0·5 M H$_2$SO$_4$	19·4 ± 0·15	2
Quinacrine, 20 mM	Water	4·0	1
Rhodamine 3GO	Water	3·9	1
Rhodamine 6G		7·5 ± 1·5	3
		5·9 ± 0·5	4
Rhodamine 6GO	Water	5·8	1
Riboflavin	Water	4·2	1
Tryptophan	Water	2·0–6·0	6

Notes:
1, Chen, Vurek & Alexander (1967); 2, Ware & Baldwin (1964); 3, Kask, Piksarv & Mets (1985); 4, Leskovar et al. (1976); 5, Arndt-Jovin et al. (1979). Neutral pH; 6, Eftink (1983)
[a] For ref. 1, unless specifically noted, the solutions contained 10 mM Tris·HCl buffer (pH 7·0). Solute concentrations were 10 μM, except that proteins were 1 mg/ml. All measurements were made at 23 °C.

than fluorescence. Mixed fluorescence and phosphorescence can occur from the same excited specimen.

The mechanism is as follows. Excitation takes place in the same way as for fluorescence. Although it is rare for molecules to enter an excited triplet state directly from the ground state, in many molecules there is an efficient process whereby an excited singlet state may be converted to an excited triplet state. This process is called *intersystem crossing*. It is immediately followed by vibrational relaxation whereby the molecule falls to the lowest vibrational energy level of the triplet excited state (T_1). If from this triplet state it returns to the ground state by emission of a photon, this is called phosphorescence. Since it involves a triplet–singlet transition, it is less probable than fluorescence and therefore takes longer. The lifetime of phosphorescence is in the range

$100\,\mu\text{s}$ to $1\,\text{s}\,(10^{-4}$ to $1\,\text{s})$ rather than the 1 to $10\,\text{ns}\,(10^{-9}$ to $10^{-8}\,\text{s})$ for fluorescence. The energy level of a triplet state is, in general, lower than that of the corresponding singlet state, so that the lowest-energy triplet excited state (T_1) is at a lower level than the lowest-energy singlet excited state (S_1). It follows that the quantum energy available for phosphorescence is generally less than that for fluorescence, and the wavelength of phosphorescence is correspondingly longer.

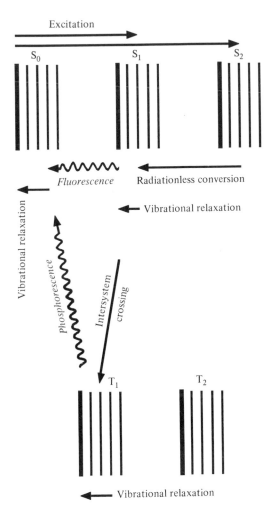

Fig. 12.1. An illustration of the transitions between molecular energy levels during absorption of light, fluorescence and phosphorescence. Only electronic and vibrational energy levels are shown. Absorption of a photon produces excitation (left) from the ground state (S_0) to an excited state $(S_1$ or $S_2)$. From S_2, S_1 is quickly reached by radiationless conversion. The energy level falls to the lowest vibrational energy level of S_1 by vibrational relaxation. Fluorescence is produced by radiative transition to S_0, with subsequent vibrational relaxation to the lowest vibrational level of the ground state. If intersystem crossing occurs from S_1 (singlet) to T_1 (triplet), phosphorescence may occur by radiative transition to S_0.

Phosphorescence can be distinguished from fluorescence by the longer delay time of the luminescence. If a photoluminescent substance is excited by a brief pulse of light this will be followed by a corresponding pulse of fluorescence or phosphorescence. Whereas the fluorescence is completed almost instantaneously, the phosphorescence may persist for some seconds or even longer. Phosphorescence can be distinguished in the presence of fluorescence by time-resolution, using flash excitation and observing after the fluorescence lifetime. Because of the relatively long times involved, this can be arranged by a system of rotating mechanical shutters.

To examine phosphorescence, some mechanical or other system is required whereby the specimen is excited by pulses of light, and examined between the pulses. This can for example be carried out by means of rotating shutters, one each in the excitation pathway and emission pathway, so arranged that the emission pathway is blocked while the specimen is illuminated and vice versa.

It is therefore possible to observe phosphorescence by modifying a fluorescence microscope with a mechanical shutter arrangement consisting of two shutters, so arranged that the excitation shutter provides a series of very brief pulses of excitation, and the emission shutter is synchronized to blank out the viewing pathway during each period of excitation. No barrier filter is necessary because the choppers completely cut off the exciting light. I have seen only one phosphorescence microscope, that devised by Professor R. Barer at Sheffield. This employed rotating shutters, whose phase could be varied; if my memory serves me correctly, the phosphorescence of Eosin was demonstrated to a small group of us during a break in a Royal Microscopical Society meeting, about 1965. Now that time-resolution of luminescence has become technically easier, perhaps phosphorescence microscopy may become more widely used.

Phosphorescence studies are best carried out at very low temperatures, e.g. in liquid nitrogen at 77 K, since at such temperatures the radiationless processes which compete with phosphorescence and fluorescence are largely suppressed, thereby greatly increasing the intensity of phosphorescence. Accordingly microphosphorometry requires provision for cooling the specimen. Instrumentation and techniques for low temperature microscopy are mentioned in Chapter 8.

Parker (1969b) described a microspectrophosphorometer, in which a fluorescence microscope and a conventional spectrophosphorometer were combined. Mechanical choppers (shutters) enabled the determination of either phosphorescence alone (with the choppers out of phase) or fluorescence plus phosphorescence (with the choppers in phase). The device was used for the identification of microdroplets of oil ingested by marine organisms from oil spillages.

There do not appear to have been any other practical applications of microphosphorometry so far. The stratum corneum of the skin phosphoresces. The phosphorescence of some fluorochromes has been studied in solution. Acridine Orange bound to DNA shows a peak absorption at about 500 nm, and delayed emissions at about 620 nm and 530 nm (see Motoda & Kubota, 1979). Phosphorescence of protein-bound Eosin and Erythrosin were studied by Garland & Moore (1979); they found delayed emission at about 680 nm, with a secondary peak at about 550–560 nm (the latter believed to be due to delayed fluorescence, see below).

Delayed fluorescence is a form of luminescence which takes place over a time scale similar to that of phosphorescence but otherwise has the nature of fluorescence. The delay is due to a double intersystem crossing, from singlet to triplet and back to singlet. The spectrum of delayed fluorescence may be different (see Grzywacz, 1967).

Delayed fluorescence and phosphorescence of plant pigments were reviewed by Krasnovsky (1982). Delayed luminescence, with spectra similar to those of the pigment fluorescence, has been observed in both aerobic and anaerobic solutions of chlorophylls and their analogues. However, the mechanism of the afterglow is different in aerobic and anaerobic conditions.

Fluorescence polarization

Light re-emitted from an excited molecule tends to be polarized in the same plane as was the absorbed quantum. More precisely, molecules preferentially absorb and emit photons of which the electric vector is parallel to the electric field produced by the oscillating electrons. Fluorescence will tend to show polarization either if the excitation beam is plane polarized or if the fluorophore molecules are dichroic, i.e. preferentially absorb light polarized in a particular plane. If the molecules are oriented in the material, the fluorescence will be partially polarized even if the excitation was unpolarized. In biology, this phenomenon is generally known as difluorescence (Zeigenspeck, 1949) by analogy with the term dichroism. Polarization of fluorescence can be detected by placing a piece of Polaroid^{TM} film as an analyser in the eyepiece of the microscope; the analyser and the specimen should be free to be rotated in order to be set to the correct orientation in relation to the polarized light.

If required, excitation by substantially plane-polarized light is normally obtained by the elimination from the exciting light of light waves whose electric vectors do not all lie in a single chosen plane, using a polarizing filter. The polarized radiation so obtained selectively excites those molecules whose absorption transition moments have a significant component in the plane of the electric vector of the exciting beam (e.g. molecules whose long axes lie essentially in the plane of polarization). If these molecules were firmly fixed, when excited with polarized light they would emit radiation which is polarized in the same direction as the exciting light. However, molecules not firmly fixed may rotate during the interval between absorption and emission, so that the emitted light will tend to be more randomly polarized. The degree to which the emitted fluorescence remains plane polarized will depend inversely upon the amount of Brownian rotation occurring in the interval between absorption and emission of light (see Chen, 1973). The rate of rotation is least for large molecules, such as phycobiliproteins.

For most dyes in solution, the relaxation time of the rotation of the molecules in solution is much shorter than the fluorescence lifetime, and therefore almost completely depolarized fluorescence is to be expected and is indeed observed. On the other hand, fluorescent macromolecules such as proteins have longer rotational relaxation times and the degree of polarization of fluorescence is indicative of the relaxation time.

Fluorescence polarization can be studied in two ways: by observation of polarized fluorescence (difluorescence) under non-polarized excitation, and measurement of the amount of depolarization after excitation with polarized light. Observations of difluorescence under non-polarized excitation are equivalent to observation of dichroism by transmitted light, and probably most commonly reflect the existence of arrays of rod-like molecules with their long axes more or less parallel. Measurements of fluorescence depolarization are normally concerned with the mobility of the fluorescent molecules. In general terms, if polarized light is used for excitation, the emitted light would be depolarized to the extent that the molecules are free to rotate during the period between excitation and emission.

Polarization fluorescence microscopy presents a number of technical problems, and has not so far been widely used. For information on normal polarization microscopy, see Patzelt (1985), Ganse (1977), James (1976), Missmahl (1966) and Oster (1955). The conversion of a normal fluorescence microscope for polarization studies requires the addition of an analyser in the observation system, and possibly a polarizer in the excitation system; the latter is not always required. Insertion of an analyser presents no particular problem; it is commonly placed in the eyepiece, and can thereby be rotated. A straight monocular body is preferable in this context, if only to avoid complications due to the beamsplitter prisms in a binocular head. If an epi-illuminator is used, containing a dichromatic mirror or other plate beamsplitter, this will produce some polarization, the extent depending on how closely the Brewster angle of the beamsplitter approaches the 45° angle at which it is set. Because the orientation of this polarization is determined by the geometry of the beamsplitter, the polarizer and analyser must be oriented correspondingly. Under these circumstances, neither the analyser nor the polarizer can appropriately be rotated, at least for quantitative studies. Therefore, probably a metallurgical microscope with a prism illuminator in part of the objective aperture should be adapted. A fluorescence epi-illuminator is made for some Leitz polarizing microscopes; see also Eisert & Beisker (1980). Alternatively, a substage condenser can be used, thereby avoiding the problems of the epi-illumination mirror. A centrable rotating stage is required to rotate the specimen.

The production of plane-polarized ultraviolet or blue light for fluorescence excitation at the intensities normally required for fluorescence microscopy requires a polarizing device relatively resistant to high-density illumination. Suitable Nicol prisms of large size are available, unfortunately at rather high cost. In my experience Polaroid[TM] is not normally suitable for this service because it cannot withstand strong irradiation, but it has been used by dos Remedios, Millikan & Morales (1972).

Micropolarimetry of fluorescence

Microfluorometers adapted for measurement of fluorescence polarization were described by MacInnes & Uretz (1966) and dos Remedios, Millikan & Morales (1972). In the instrument of MacInnes & Uretz, the excitation (using dia-illumination) was unpolarized. The emission passed through a rotating Polaroid[TM] polarizing filter and thence to a photomultiplier, whose output was recorded on a chart recorder. When measuring fluorescence from aligned material, an oscillating trace was recorded. The percentage of polarized emission was calculated as 100 times the difference between maximum and minimum intensity values divided by the maximum value, after correction of the observed maximum and minimum values for background. If the background fluorescence is substantially independent of the analyser position, the formula for the percentage polarization (R) reduces to:

$$R = 100 \times \frac{F_{max} - F_{min}}{F_{max} - F_{background}}$$

where F_{max} and F_{min} are the maximum and minimum readings and $F_{background}$ is the background reading.

Another micropolarimeter was described by dos Remedios, Millikan & Morales (1972) and dos Remedios, Yount & Morales (1972) for rapid measurement of polarization of tryptophan fluorescence from single muscle fibres during relaxation, contraction, and rigor. Using dia-illumination through a quartz condenser, the specimen was illuminated with polarized light at 300 nm. Fluorescence was filtered to select the wavelength band 340–450 nm, and then split into two orthogonally polarized components by a Wollaston prism; the two components were measured simultaneously by photomultipliers, and the polarization of fluorescence calculated by computer. In conjunction with appropriate equipment for isolated muscle fibres, the instrument enables the collection of information about the orientation and changes in orientation of mysosin crossbridges in living muscle fibres in various physiological states (dos Remedios, Millikan & Morales, 1972; dos Remedios, Yount & Morales, 1972).

Difluorescence in animal tissues

Analysis of the polarization of fluorescence of Acridine Orange bound to *Drosophila virilis* chromosomes suggested that the DNA of interband regions of these chromosomes must lie parallel to the chromosomal axis (MacInnes & Uretz, 1966). Probably difluorescence microscopy could be applied to amyloid stained with Congo Red and possibly other fluorochromes.

Difluorescence in plants

Polarized autofluorescence has been demonstrated in chlorophyll in intact plastids (Ruch, 1957, 1966c; Olson, Jennings & Butler, 1964; Geacintov, Van Nostrand & Becker, 1974) and in lignified cell walls (Hengartner, 1961; Frey-Wyssling, 1964; Ruch, 1966a). Difluorescence is also shown by cell walls following staining with a wide variety of fluorochromes (Zeigenspeck, 1949). In these cases, the polarization is believed to be due to orientation of the fluorescing molecules bound to microtextural features (membranes in chloroplasts and fibrillar components of cell walls). Lignin itself is believed not to contribute to difluorescence, being in an amorphous state in the cell wall (Ruch, 1966c).

Difluorescence in materials science

Combined difluorescence and dichroism microscopy has been applied to cotton fibres under stress (Isings, 1966). Difluorescence microscopy offers many possibilities for petrological studies, particularly using the addition of a fluorescence system to a standard metallurgical microscope.

13

Quantitative enzyme studies

The demonstration of enzyme activity by fluorescence techniques is discussed in *Fluorescence microscopy* (Rost, 1990). The present chapter deals with quantitative aspects of the study of enzyme systems by microfluorometry. The subject has two aspects: first, the determination of the amount of enzyme activity present in a given histological or cytological region; and second, the determination, by cytochemical means, of kinetic characteristics of the enzyme, such as the Michaelis–Menten constant (K_m) and response to activators and inhibitors. A previous brief review is that of Rost (1971).

Enzyme activity can be quantified and characterized in cells or tissues using three methods: biochemical, cytochemical, and flow cytometric. In many cases, the same enzymatic reaction can be carried out under any or all of the three conditions. Biochemical assay on homogenized tissue samples enables precise quantification; however, the processes of homogenization and extraction destroy the cell as a functional unit and there may be doubt that the enzyme system may be damaged in the process of obtaining the homogenate (Rost, Bollman & Moss, 1973; Malin-Berdel & Valet, 1980), and the results may be difficult to interpret on a cellular level, since more than one cell type is invariably present. Cytochemical quantification enables the cells to be kept in the tissue and the exact localization of each measurement is known. Flow cytometry combines many of the advantages of the biochemical and cytochemical methods, and also permits quantification in each cell of other parameters such as cell volume.

Microfluorometric techniques have been developed for the study *in situ* of some oxidative enzyme systems (Chance & Legallais, 1959; Lowry & Passoneau, 1972; Kohen *et al.*, 1973), glucose-6-phosphate dehydrogenase (Täljedal, 1970; Galjaard *et al.*, 1974; Jongkind *et al.*, 1974); and some hydrolytic enzymes, namely non-specific carboxyesterase (EC 3.1.1.1; Rotman & Papermaster, 1966; Sernetz & Thaer, 1973), alkaline phosphatase (EC 3.1.3.1; Rost, Bollmann & Moss, 1973), glucosidase and galactosidase (Galjaard *et al.*, 1974), and acid phosphatase (EC 3.1.3.2; Prenna, Bottiroli & Mazzini, 1977).

The same microfluorometric equipment can also be adapted to biochemical measurement of very small volumes, e.g. in microdroplets (Jongkind *et al.*, 1974) or using the aperture-defined microvolume (ADM) method (for a review, see Tanke *et al.*, 1985).

Microfluorometric quantification of enzyme activity

Methodology

The methodology for quantification of enzyme activity in a particular cytochemical region is essentially similar to that employing microdensitometry: the tissue is incubated for a fixed period of time in an appropriate solution and the amount of final reaction product (FRP) generated is measured by microfluorometry and compared with a standard.

Hydrolytic enzymes

For the study of hydrolytic enzymes, artificial substrates are employed in which the substrate is a relatively non-fluorescent ester of a strongly fluorescent substance, such as Naphthol AS-BI or in some cases a more soluble substance such as fluorescein, 1-naphthol, 2-naphthol, or 4-methylumbelliferone.

Oxidoreductases

For the study of oxidoreductases, the fluorescence of reduced NADH is monitored. NADH (reduced nicotinamide adenine dinucleotide, I) bound to a mitochondrial component or to a dehydrogenase is believed to be the main source of blue autofluorescence within cells (Chance & Thorell, 1959a; Aubin, 1979). The fluorescence has excitation/emission maxima at about 365/445 nm (Aubin, 1979); fluorescence microscopy requires UV excitation, and a colourless barrier filter. Only the reduced form (NADH) is fluorescent; the oxidized form (NAD$^+$, II) is non-fluorescent. There is

I NADH

II NAD$^+$

accordingly a change in cellular autofluorescence according to the redox conditions obtaining within the cell. This can be measured by microfluorometry (Chance & Thorell, 1959a,b). This subject is discussed further below.

Accuracy of enzyme assays

Enzyme assay by histochemical methods involves three stages: demonstration of the enzyme; measurement of the optical density or fluorescence intensity of the final reaction product; and the process of relating the numerical result obtained to the actual amount of enzyme activity present per topographic unit (i.e. activity per cell, per unit weight, or per organelle).

A significant cause of error is loss of enzyme activity, which can occur in several ways, of which the most important are believed to be, firstly, actual loss of the enzyme by diffusion, and secondly, inactivation of the enzyme by the fixative, subsequent tissue processing, and substances present in the incubation mixture.

The amount of loss by diffusion varies for each enzyme, and for each fixative. One function of the fixative is of course to prevent diffusion of the enzyme into the incubation mixture, but unfortunately some diffusion invariably takes place into the fixative itself, and moreover the process of fixation results in some inactivation of the enzyme (see Burstone, 1958; Nachlas, Young & Seligman, 1957; Scarpelli & Pearse, 1958; Seligman, Chauncey & Nachlas, 1951). In practice, a compromise may have to be reached between adequate fixation for accurate localization, and a lesser degree of fixation with preservation of enzyme activity. Fixation notwithstanding, some diffusion of enzyme will take place into the incubation medium, the extent of this loss being influenced by the duration of incubation, temperature, pH, the electrolyte concentration, and substances which may be added to the incubation medium specially to reduce enzyme diffusion.

Measurements on tissue sections, unless used for comparison with a single section, are dependent upon knowledge of the section thickness. There does not appear to be any really satisfactory method at present for measuring section thickness; interferometry is probably best. Accuracy is also affected by instrumental error in measurement of optical density or fluorescence. This is a purely technical problem (see Chapter 4) and is probably the least significant source of error in this context.

Obviously the final evaluation of any quantitative histochemical system will depend upon comparison with results obtained by such other procedures as may be available, and upon the results of calibration by proper model systems. For example, model systems based on the incorporation of alkaline or acid phosphatase into polyacrylamide gels have been devised by van Duijn and his coworkers (van Duijn, Pascoes & van der Ploeg, 1967; Lojda, van der Ploeg & van Duijn, 1967; van der Ploeg & van Duijn, 1968).

Ideally all cytochemical techniques for the localization and assay of enzymes in tissue sections or single cells should be *correctly localized*, *reproducible*, *specific* and *valid* (Stoward, 1980). A *correctly localized* technique may be defined as one in which the FRP is deposited only at or on the true *in vivo* intracellular site of the enzyme, and nowhere else. Stoward (1980) used the term *precise* rather than *correctly localized*; however, the word 'precise' has a specific meaning in respect of the final numerical result and I prefer the term 'correct localization' even if it is longer and rather clumsy. Perhaps there is yet a better term?

A *reproducible* technique gives essentially the same results when repeated, provided that exactly the same procedure is followed each time and that the biological samples (tissues or cells) are equivalent.

A *specific* technique is one which demonstrates or measures on the enzyme in question or, if several FRPs are formed, gives rise to a component of the FRP that can be clearly identified and attributed to the activity of the enzyme.

A *valid* technique can be roughly defined as one in which the amount of FRP deposited per unit volume (or other reference parameter) in a particular region of a cell (i.e. the intensity of fluorescence) is related to either the concentration or the specific activity of the enzyme in that region.

Stoward (1980) proposed that a technique can be considered to be correctly localized, reproducible, specific and valid if it meets the following practical criteria:

1. Precise localization:
    (i) Sections retain their morphology and look 'clean'.
    (ii) Specific FRP is confined to certain subcellular sites, particularly those predicted to contain the enzyme on the basis of other experiments.
    (iii) There is positive proof that specific FRP does not diffuse and bind to other subcellular sites.

2. Reproducibility:
    (i) The mean values of measurable parameters (e.g. the mean fluorescence emission of FRP) do not vary significantly in repeated experiments.
    (ii) The individual measurements of these parameters within a preparation (e.g. the absorbance of the FRP at its absorption maximum) statistically form a unimodal population.

3. Specificity:
    (i) No specific FRP is formed in control sections or preparations.
    (ii) The reaction conditions (pH, etc.) that give rise to the maximum rate of formation of specific FRP *in situ* is the same as, or very similar to, those favouring the optimal formation of reaction product of systems *in vitro*.
    (iii) Inhibitors and other enzyme modifiers exert their expected effects on the formation of specific FRP in accordance with 'biochemical' precedents.
    (iv) Potentially interfering enzyme systems have either been suppressed or shown to be absent, or can be distinguished from the enzyme under study.

4. Validity:
    (i) No enzyme is lost from its subcellular site during the procedures required for its visualization, or, if some enzyme is inevitably lost, the loss is small (say less than 30%), constant and known.

(ii)   The specific FRP arising from an enzyme reaction:
(a) has its expected chemical composition, and
(b) can be identified with reasonable certainty in cells and tissues (by, for example, having fluorescence excitation and emission maxima at wavelengths similar to that of pure specific FRP in solution or bound to a non-specific model carrier), and
(c) has a stoichiometric relationship with the amount of primary reaction product (PRP) formed by the enzyme.

(iii) The mean absorbance or fluorescence emission of the specific FRP is proportional to its mean concentration in the cell or section.

(iv)   For sections incubated for a constant time in media containing an excess of substrate, the mean fluorescence emission of the specific FRP in cells of the same histological type is proportional to the thickness of the section up to a certain 'critical' level. Stoward (1980) proposed absorbance, rather than fluorescence emission, in respect of this criterion, adding that it might be unrealistic to expect such a proportionality of fluorescence, because of reabsorption of fluorescence emission. Obviously this latter would depend in part on whether or not epi-illumination is employed.

(v)   The rate of increase of mean fluorescence emission per unit incubation time, i.e. apparent reaction rate, in a particular site is directly proportional to the specific activity or concentration of the enzyme in that site.

(vi)   Once specific FRP has begun to form in a cell, its mean fluorescence emission increases uniformly with incubation time, preferably linearly.

(vii) On extrapolated plots of fluorescence versus incubation time, the mean emission of the specific FRP corresponding to zero incubation time should ideally be zero, or, at worst, small but constant for a particular set of reagents and reaction conditions.

Certain enzymes, particularly those thought to be contained within membrane-bound organelles, may show a time lag before FRP begins to be formed.

(viii) The rate of formation of specific FRP in whole cells or sections of constant thickness is a function of the concentration of substrate in the incubation medium. This means that at low substrate concentration the amount of FRP formed per unit incubation time increases as the concentration increases, but when the substrate concentration is above a certain level, the reaction rate reaches a constant maximum.

(ix)   Reciprocal or other suitable plots relating substrate concentrations and the observed reaction rates of an enzyme *in situ* should yield Michaelis-Menten constants (K_m) comparable to, or possibly higher than, that obtained for the same enzyme at substrates in solution.

(x)   The changes in the rates at which specific FRP is formed *in situ* in the presence of low concentrations of enzyme modifiers (especially inhibitors) should be of an order comparable to those exhibited by the enzyme *in vitro*.

(xi)   In control sections and preparations of either fresh samples incubated with media lacking an essential ingredient (such as substrate) or boiled samples incubated with complete media, the mean fluorescence of any non-specific FRP is small and preferably constant, say not more than about 5–10% of the mean fluorescence of the specific FRP.

In view of the difficulties and the limitations which have been discovered in trying to apply rigidly the criteria listed above, together with the likelihood that most 'applied' investigators may have neither the inclination nor the facilities to carry out all the tests which these criteria imply, Stoward (1980) suggested ten working guidelines for quantitative microdensitometric technique. It is not difficult to translate these criteria into microfluorometric terms. One additional criterion is required: that the FRP be not significantly subject to fading under irradiation during microfluorometry. Criteria for microfluorometric study of enzyme activity have also been discussed by Prenna, Mazzini & Cova (1974) and Prenna, Bottiroli & Mazzini (1977), mainly in relation to acid phosphatase with Naphthol AS-BI phosphate as substrate.

On the basis of the above, the following working criteria are suggested for the microfluorometric assay of enzyme activity. If a procedure complies with the following eleven guidelines, or at the very least the first six, then it may be expected to be sufficiently valid for most studies requiring the determination of relative values of enzyme activity. The eleven suggested criteria are:

1. There is adequate qualitative evidence for believing that the cytochemical reaction on which the technique is based is reasonably correctly localized and specific.

2. The composition of the reaction medium (substrate concentration, pH, etc.) is such that the enzyme produces FRP at a maximal rate (V_{max}) and this rate is unaffected by small changes in the composition of the incubation medium.

3. Once FRP has begun to be formed its fluorescence emission increases steadily with incubation time, preferably linearly.

4. In model sections containing pure enzyme the rate at which the fluorescence emission of the specific FRP increases (linearly) per unit incubation time is proportional to the relative concentration of the enzyme.

5. No enzyme is lost from the preparations during the procedure, or, if a loss is inevitable, a constant and preferably small proportion of the original enzyme content diffuses into the incubation medium.

6. There is no significant photodecomposition of the FRP during irradiation.

7. The thicknesses of sections used in histochemical assays are constant, and less than about 14 μm (this restriction does not apply to intact single cells).

8. The concentration of the FRP is kept low enough to avoid 'concentration quenching', i.e. error due to reabsorption of emitted light.

9. The fluorescence corresponding to zero incubation time is zero or small in comparison to that at the end of incubation.

10. The rate at which non-specific FRP is formed in control preparations is low (Stoward (1980) suggests less than 30% of the rate at which specific FRP is deposited).

11. The rates of increase of fluorescence of specific FRP per unit incubation time (i.e. reaction rates) are reproducible (i.e. show no significant differences statistically) in preparations from at least four different samples of similar physiological status.

Enzyme kinetics

Fluorometry is particularly suitable for kinetic studies in living cells or unfixed sections, because of the high sensitivity of microfluorometry. The instrumentation required consists essentially of a standard microfluorometer, modified to include a heated stage with thermoregulation to maintain the desired temperature, an incubation chamber on the stage of the microfluorometer, and a data recording system which will record changes in fluorescence intensity with time: this latter can vary from a strip chart recorder to a computerized system taking readings at preset intervals of time. Since the fluorescence is monitored during the enzymatic reaction, the deleterious effects of ultraviolet (UV) radiation must be minimized by using the lowest possible radiation (in intensity and duration).

Oxidative enzymes

Kinetic studies of oxidative enzyme systems in living cells have been made by microfluorometry, utilizing the fluorescence of reduced pyridine nucleotide (NADH) referred to above. Pioneering studies were described by Chance & Thorell (1959a,b), Chance (1962), Kohen (1964a), Kohen & Kohen (1966), and Kohen, Kohen & Thorell (1968a,b, 1969). There is now a substantial literature on this subject. For reviews, see Kohen *et al.* (1987, 1983, 1981a,b, 1969), Kohen, Hirschberg & Rabinovitch (1985), and Salmon *et al.* (1982). See also Kucera & Deribaupierre (1980). Microfluorometry of fluorescent cellular coenzymes, namely NAD(P)H and flavins, in conjunction with sequential microinjections of metabolites and metabolic modifiers into the same cell, has produced valuable data on the regulatory mechanisms of transient redox changes of mitochondrial and extra-mitochondrial redox pathways (Kohen *et al.*, 1981b, 1983, 1985). The possibility of measuring from intracellular regions as small as a few micrometres in diameter allows correlation of cytological structure and biochemical function (Kohen, 1964b; Kohen *et al.*, 1983, 1985).

Instrumentation and techniques were reviewed by Kohen *et al.* (1981a,b). The specimen must be isolated from atmospheric oxygen and kept in an atmosphere which does not affect the redox state of the tissue, otherwise the fluorescent NADH formed is quickly converted to NAD^+. Substrates and other substances of interest can be microinjected into a cell (Nastuk, 1953; Kopac, 1964; Kohen & Legallais, 1965; Kohen, Kohen & Jenkins, 1966; Kohen, Legallais & Kohen, 1966; Meech, 1981). Chance,

Thorell, Kohen and co-workers used a differential microfluorometer (Chance & Legallais, 1959; Kohen et al., 1967, 1969) which was subsequently equipped with a flowchamber (Kohen, 1963) and apparatus for microinjection. Later studies were made with the instrument of Hirschberg et al. (1979).

Täljedal (1970) studied the apparent K_m of glucose-6-phosphate dehydrogenase in cryostat sections by microfluorometry of NADPH.

NADH fluorescence has also been measured from the surfaces of a number of organs, namely cardiac muscle (Chance et al., 1965; Williamson & Jamieson, 1965; Moravec et al., 1972; Nuutinen, 1984; Koretsky, Katz & Balaban, 1987), liver (Ji et al., 1982), and kidney (Franke, Barlow & Chance, 1980). Such studies can conveniently be carried out with microfluorometric equipment, slightly modified to accept the specimen. It is important to note that the specimen must be kept in a non-oxidizing atmosphere; atmospheric oxygen must be barred from the system, otherwise the NADH on the surface is promptly oxidized to NAD^+ and the fluorescence disappears. Nitrogen is a suitable medium to fill a chamber containing the specimen; the fluorescence can be observed through a window in the chamber.

Microspectrofluorometry of flavins (Chance et al., 1968) presents a number of difficulties and is less far advanced. Not all flavins are fluorescent, and the degree of participation of flavins in metabolic processes appears to be variable (Kohen et al., 1985).

Cytochemical methods for the study of oxidoreductases in tissues are more usually based on the reduction of tetrazolium salts to insoluble formazans, which form a precipitate at the site of enzyme action. This method has been adapted to fluorescence microscopy by the introduction of cyanotoluyltetrazolium chloride (CTC) which gives rise to a fluorescent formazan, which can be used for both morphological studies (Stellmach, 1984) and kinetic studies (Severin & Stellmach, 1984; Severin, Stellmach & Nachtigal, 1985). The formazan fluoresces red, and requires green excitation.

Hydrolytic enzymes

Rotman & Papermaster (1966) observed that living cells in a medium containing certain fluorescein esters of fatty acids became fluorescent. This was due to the non-polar ester passing through the cell membrane and becoming hydrolyzed by intracellular esterases to produce free fluorescein, which, being polar, could not readily pass the cell membrane and therefore accumulated in the cell. This phenomenon allows kinetic investigations of the properties of cell membranes (Rotman & Papermaster, 1966; Sernetz, personal communication, 1969; Sernetz, 1973; Boender, 1984), and is the basis of a widely used test for viability of cells in animals (Rotman & Papermaster, 1966; Poel et al., 1981; Hutz, DeMayo & Dunkelow, 1985) and plants (Widholm, 1972; Heslop-Harrison & Heslop-Harrison, 1970).

The use of cryostat sections of isolated living cells for enzyme kinetic studies has several potential advantages compared with the use of enzyme solutions. For example, alkaline phosphatase can be obtained free in solution only by vigorous procedures such as autolysis or extraction with n-butanol, since it is firmly bound to structural elements of the cells in which it occurs. These extraction procedures are potentially disruptive of enzyme structure and therefore it may be doubted whether characteristics of the enzyme solution such as substrate specificity, K_m, pH optimum, or response to activators or inhibitors truly reflect the properties of the enzyme within the living cell. We (Rost, Nägel & Moss, 1970) reported preliminary experiments using α-naphthyl orthophosphatase (Moss, 1960), which was studied in cryostat sections mounted in the flow chamber of the MRC microspectrofluorometer (Pearse & Rost, 1969; Rost, 1973); we obtained an apparent K_m value and measures of pH-dependence in good agreement with the corresponding values for the soluble enzyme. We concluded that the extractive procedures used in biochemistry did not significantly affect the enzyme; while the good agreement of the biochemical and cytochemical data gave mutual support for the validity of both. Acid phosphatase has been studied in a similar manner by Prenna *et al.* (1977), using naphthol AS-BI phosphate, and by de Permentier & Rost (unpublished) using 1-naphthol phosphate.

Instrumentation

The instrumentation which my colleagues and I have developed for studies of hydrolytic enzyme kinetics (Figs. 13.1, 13.2) may be of interest. Cytochemical studies

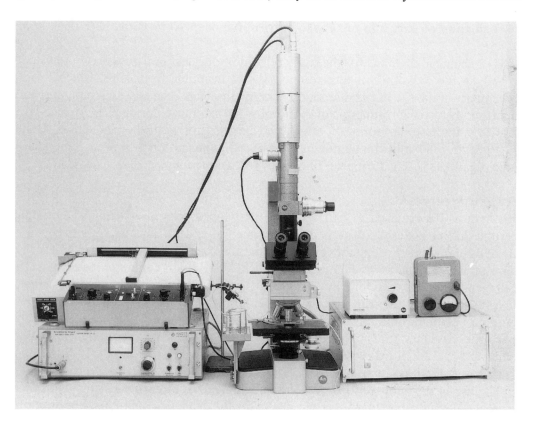

Fig. 13.1. A microfluorometer for studies of enzyme kinetics, based on a Leitz Orthoplan microscope with temperature-controlled stage and a Leitz MPV-1 photometer. The microscope has been modified by the attachment of a platform to accommodate the beaker of substrate solution. From de Permentier (1981).

are carried out using a microfluorometer (Fig. 13.1) based on a Leitz Orthoplan microscope and a Leitz MPV-1 photometer, both slightly modified. The light source is a 200 W high-pressure mercury arc lamp (HBO 200), with a stabilized power supply. A PloemOpak II epi-illuminator is used; Block A (excitation filters UG1 and BG38, dichromatic mirror TK 400 and barrier filter K430 provides UV excitation). A Zernike phase-contrast system, the condenser having a long working distance (11 mm), is used for locating the required area of the section.

The MPV-1 photometer is mounted on top of the vertical phototube of the microscope. The usual measuring diaphragm of the MPV-1 is replaced by a fixed circular aperture, corresponding to a circular measuring field 150 μm in diameter. For visual examination of the measured field, the Leica Visoflex of the MPV-1 is replaced by a tube fitted with an ocular. For photometry, the emission wavelength is selected by an interference filter. The photomultiplier tube is connected directly to a chart recorder, run at a speed of 60 mm/min.

The incubation chamber (Fig. 13.2) for histochemical studies is similar to that previously described (Rost, Bollman & Moss, 1973). The chamber is made by attaching a plastic (methacrylate) ring to a glass slide; two narrow plastic tubes are glued into holes on opposite sides of the ring. The chamber is completed by a glass coverslip, carrying the tissue section underneath. One of the tubes (the inlet tube) is attached to a two-way tap from which two tubes lead to buffer and substrate solution, respectively. The other tube (the outlet tube) is attached to a syringe; sucking with the syringe draws

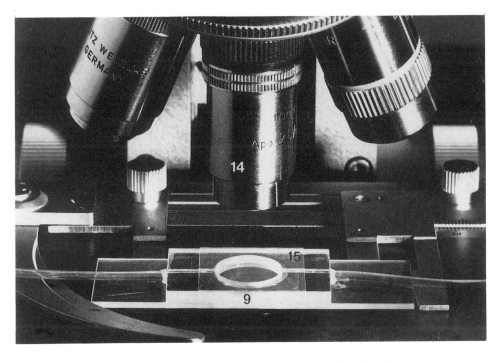

Fig. 13.2 A close-up view of the incubation chamber when placed on the stage of the microfluorometer. Key: 9, incubation chamber; 14, two-way tap; 15, coverslip with specimen underneath. From de Permentier (1981).

fluid into the chamber. The temperature inside the chamber is measured by a thermistor, the resistance of which is measured to 0·001 ohm with a digital ohmmeter, and calibrated against a standard thermometer reading to 0·1 Celsius degree. Following the calibration, it was found that a regression equation could be fitted very closely, and this equation is used for conversions between resistance readings and temperature. The temperature of the incubation chamber is maintained at the desired level by the use of a heating and cooling stage (Leitz, Model 80) heated by a built-in electric element and cooled by a Leitz Kryomat refrigerator unit. The supplies of substrate solutions and buffer, waiting to be introduced into the chamber, must be maintained at the required temperature.

Histochemical studies are carried out in a manner similar to that already described (Rost, Bollman & Moss, 1973). Briefly, a coverslip carrying a tissue section is placed on the incubation chamber, with the section facing downwards into the chamber. With the chamber filled initially with buffer, an appropriate area of the section is selected. The required substrate solution is then drawn into the chamber. When a sufficient length of graph has been accumulated on the chart recorder, the incubation solution is washed out with buffer and a new solution drawn in. The slope of the graph (fluorescence intensity versus time) gives the relative velocity of the reaction; plotting relative velocity against substrate concentration allows calculation of the Michaelis–Menten constant (see Wilkinson (1961) for calculations).

14

Flow cytometry

Hans J. Tanke

Department of Cytochemistry and Cytometry, University of Leiden, Netherlands

The principle of flow cytometry is that cells or cellular components in aqueous suspension are passed through a sensing region where optical or electrical signals are generated and measured. The typical analysis rate of commercial instruments is in the order of several thousand objects per second. Cells are generally fluorescently stained, although non-fluorescing dyes in sufficient concentration can be measured as well on the basis of axial light absorption. Staining is not required for measurements of cell size by light scatter or electrical resistance. The reproducibility of fluorescence measurements is 2% or better, and the detection limit of most commercial instruments is 2000–3000 molecules of fluorescein per cell.

A typical configuration of a flow cytometer is shown in Fig. 14.1. The cells are hydrodynamically forced with a constant speed of 5–10 m/s to a region onto which a high-intensity light source, usually a laser or high-pressure arc lamp, is focussed to generate light-scatter signals and fluorescence emission. Properly oriented photodetectors collect a fraction of the signals and generate electrical signals proportional to the optical signals. These signals are accumulated, analysed and usually presented as single or multi-parameter frequency distributions (histograms). Manual interpretation or mathematical analysis of these histograms provides biological information of the measured cell population. Operation of modern instruments and interpretation of the data are highly facilitated by the use of microcomputers, by which also a large degree of automation of the whole procedure has been achieved.

Typical applications of flow cytometry include the analysis of DNA content in cells for clinical pathology, the analysis of immunocytochemically stained membrane antigens in immunology, and the counting of various blood cell types in haematology.

Besides analysis, several flow cytometers allow the sorting of cells of interest on the basis of the recorded signals. For this purpose acoustic energy is transferred to the fluid causing the jet to break up into uniform equally spaced droplets after the laser intersection point. These droplets containing the cells of interest may be charged depending on the signals measured, and deflected from the main stream in an electric field (Fig. 14.1). This allows the isolation of (live) biological objects with defined properties on the basis of measurements of light scatter or fluorescence emission for cell culture, biochemical analysis, or for visual examination under the microscope.

As discussed previously, a variety of instruments has been developed based on different principles, but with the same purpose: the acquisition of quantitative information from cells in flow. Consequently, various names for this technique have been used, such as flow cytofluorometry, pulse-cytophotometry or flow cytophotometry. The commonly accepted name 'flow cytometry' will be used throughout this chapter.

A flow cytometric procedure basically consists of three steps: (1) isolation, preparation and staining of the material, (2) the measuring procedure itself (acquisition of cytometric data such as light scatter, fluorescence, cell volume), and (3) analysis and interpretation of the data. These steps are discussed in more detail below. Major reviews of flow cytometry are those of Melamed, Lindmo & Mendelsohn (1989), Shapiro (1988) and Van Dilla *et al.* (1985).

Historical overview

Development of flow cytometric equipment started almost 50 years ago with the construction of systems aiming at the counting of particles in suspension, such as blood

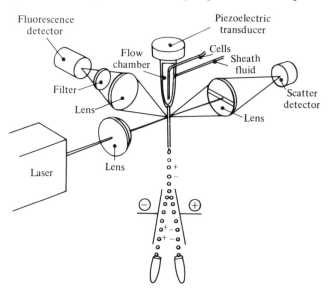

Fig. 14.1. Schematic representation of a fluorescence flow cytometer and sorter, with orthogonal axes of cell flow, illumination and emission. Fluorescently stained cells are passed through a focussed laser beam, and the generated light scatter and fluorescence emission signals are sensed, digitized and displayed as frequency distributions. Two sorting windows may be defined for each single parameter or combination of parameters. After the sensing area the fluid stream breaks up into uniform droplets generated by a piezoelectric transducer. Sorting is accomplished by selectively charging the fluid stream (positively or negatively for the two defined sorting windows) with an electrode, exactly at the moment that the cell of interest is about to pinch off into a droplet. Charged droplets are consequently deflected in an electric field downstream and collected. The example shown allows detection of forward-angle light scatter and one fluorescence parameter. Extensions may include a Coulter sensor or additional photomultipliers for measurements of multi-colour fluorescence or for perpendicular light scatter.

cells. In 1941, Kielland described a method for counting blood cells, very similar to an earlier procedure of Moldavan (1934). He used an apparatus in which cells in suspension were directed through a capillary tube on a microscope stage. The passage of each cell was registered by a photosensor attached to the ocular. Major problems of these early prototypes were the maintenance of flow in narrow channels, avoidance of obstructions by larger objects (cell aggregates) and proper focussing. A significant improvement was made in 1953 when Crosland-Taylor introduced hydrodynamic focussing for flow cytometry (Crosland-Taylor, 1953), based on pioneering studies by Reynolds in 1883 on laminar flow. Cells were injected slowly in the middle of a relatively fast-flowing sheath fluid permitting focussing of the cells within a small inner core of a larger channel. Most of the commercially available flow systems nowadays use this principle of hydrodynamic focussing. Besides photoelectric sensing as used by Moldavan, electric sensing was investigated. One of the earliest flow systems is the Coulter-type cell counter (Coulter, 1949). In this apparatus, cells are passed through an aperture where the presence of an object results in a detectable change in conductivity between two electrodes provided that cells and suspension medium differ in conductivity. Kamentski, Melamed & Derman demonstrated in 1965 that cellular macromolecules can be quantified using flow cytometry and multiparameter spectrophotometry.

Shortly thereafter, the use of fluorescence was introduced by various groups. In 1968 and 1969, Dittrich & Goehde described a microscope-based system for the measurement of particles moving in a flow stream parallel to the optical axis of a light microscope and through the focal plane of an objective with high numerical aperture (see also Fig. 14.2). Van Dilla and colleagues at the Los Alamos National Laboratory (USA) simultaneously developed so called orthogonal flow systems (Van Dilla et al., 1969). These systems had three orthogonal pathways: fluid stream, excitation pathway (lasers were introduced at this moment as excitation light sources), and the detection pathway for light (fluorescence emission, scattered light). Although the Dittrich–Goehde flow cytometer has proven to be a relatively inexpensive but sensitive and accurate measuring instrument, the orthogonal systems had a major advantage. The configuration of the latter allowed physical separation of measured particles from the main stream by means of electrostatic deflection of droplets containing the cells of interest. This principle of cell sorting was first described by Fulwyler (1965). A multiparameter flow cytometer incorporating this principle was described by Bonner et al. (1972). Cell sorting followed by visual examination of the isolated cells using a fluorescence microscope offers a way of checking the measuring data by correlating the measurements with the observed microscopic image. Moreover, sorting of cells (live or dead) or subcellular organelles such as chromosomes, on the basis of physical or immunological properties for biochemical purposes (cloning of DNA) or reconstruction experiments, offers a unique and extremely flexible tool in cell biology and experimental medicine. This chapter will not further discuss aspects and applications of cell sorting. The statement 'if it can be measured, it can be sorted' holds for most cases. A recent but very important step in development of flow cytometric instrumentation has been the introduction of microcomputers for signal handling and manipulation, and for operation control. Flow cytometers have developed from complicated research instruments towards user-friendly laboratory systems. At present, flow cytometric facilities are found in most university hospitals and are used increasingly for clinical applications.

Preparation and cytochemical staining for flow cytometry

A cell suspension of good quality is a prerequisite for optimal flow cytometric analysis. Such a suspension ideally consists of well-preserved single cells, without too much cell debris and other artefacts such as cell aggregates and clumps. This may be easily achieved for blood cells, but is very difficult in the case of solid tissues. Moreover, one general method suitable for all types of tissue cannot be given. As is important for all sampling procedures care should be taken that the end result is a true reflection of the native sample, and that no selective loss of one particular cell type occurs in case of heterogeneous organs (and most of them are!). Isolation of cells from solid tissue is generally based on mechanical disaggregation of the tissue or on enzymatic digestion (often a combination of both).

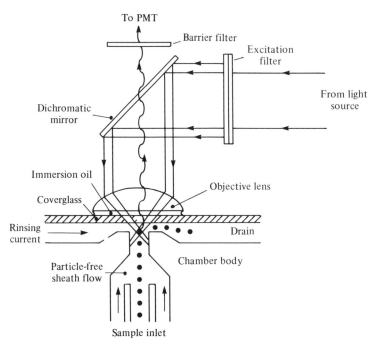

Fig. 14.2. Schematic representation of a flow-through chamber, where the fluid stream with the objects flows parallel to the optical axis. Focussing problems do not occur, since each object has to go through the optimal focus plane of the objective. The excitation is performed according to the principle of Köhler illumination. Detection optics and pulse processing are similar to those shown in Fig. 14.1. This configuration is not suitable for sorting based on deflection of droplets. However, sorting may be accomplished in a adapted flow chamber (not shown here), where the various fluid pathways after the illumination area can be closed by instantaneously generated gas bubbles (electrolysis), thereby forcing the cells of interest into a defined direction. The speed, however, is much lower than is achievable with sorters based on electrostatic deflection of droplets. PMT, photomultiplier tube.

Mechanical disaggregation may consist of simple cutting of the tissue with razor blades followed by syringing of the smaller pieces to generate more single cells. The suspension is then sieved through appropriate nylon mesh filters. Another method consists of scraping cells from a piece of tissue with the edge of a microscope glass slide. Mechanical disaggregation generally results in sufficient cells for flow analysis. However, the procedure may lead to considerable cell damage. Whether this will influence the flow cytometric analysis strongly depends on the parameters under investigation. For instance, mechanical cell damage generally will affect analysis of membrane antigens more than that of nuclear DNA.

Enzymatic digestion of fresh tissue consists, in general, of the following steps. Tissue obtained by surgery is placed in cold culture medium, supplemented with 10% (v/v) fetal calf serum. The tissue is minced and washed with medium and subsequently incubated in medium containing the digestive enzyme under gentle stirring. The cell suspension is decanted from the tissue fragments, cooled in ice and filtered through appropriate filters to obtain a suspension of single cells. Enzymes that may be used are trypsin, pronase, collagenase and hyaluronidase. The procedures discussed above may be used for fresh or frozen tissue. Recently, it has become possible to prepare formalin-fixed *paraffin-embedded archival material* for flow cytometric analysis of DNA content (Hedley *et al.*, 1983). The method consists of preparing 40–60 μm sections from the tissue blocks, which are subsequently dewaxed in xylene and rehydrated. A nuclear cell suspension may then be prepared by digesting the sections with pepsin. Most of the connective tissue as well as the cell cytoplasm is digested by this procedure. Since this procedure allows retrospective investigations, it is very useful for prognostic studies of DNA aneuploidy in solid tumours.

Fixation of the cells in suspension prior to staining may be necessary, either to guarantee preservation of the cellular components or simply to permeabilize membranes for dye accessibility. Two main aspects are important: some fixatives may induce cell aggregation (for instance ethanol), other fixatives will influence the fluorescence properties of fluorochromes (for instance glutaraldehyde and formaldehyde cause quenching of the nucleic acid dyes Ethidium and Propidium). Clumping may be kept minimal by vigorous stirring of the cell suspension while slowly increasing the ethanol concentration to preferably 70% (v/v). Fixation in paraformaldehyde (1% in phosphate-buffered saline) may be used for immunofluorescence studies. The fixative induces minor amounts of autofluorescence, and may be applied prior to immunofluorescent staining as well as afterwards without significantly influencing the results.

Cytochemical staining methods for flow cytometry fall mostly within the following categories:

1. Staining procedures for nucleic acids, sometimes in combination with general protein stains.
2. Immunocytochemical staining methods, using fluorescently labelled antibodies.
3. Enzymatic staining, using chromogenic (fluorogenic) substrates.

DNA measurements have proved to be useful for the study of cell cycle kinetics and in investigating the diagnostic and prognostic value of DNA ploidy in malignancies. Most dyes are used under so-called equilibrium conditions, e.g. cells are resuspended in dye solution, allowed to take up stain and analysed by flow cytometry while suspended in dye solution. Among the fluorescent DNA dyes some show base specificity; Chromomycin and 7-Aminoactinomycin D are G·C-specific non-intercalating dyes, whereas the non-intercalating dye bisbenzimide (Hoechst 33258) binds preferentially to A·T base pairs. Other dyes such as the phenanthridinium dyes Ethidium and Propidium intercalate in double-stranded nucleic acid without much base specificity. When used for quantification of DNA, RNase treatment is required, since binding to double-stranded RNA will occur as well. An excellent overview of fluorescent probes for DNA has been given by Latt (1979).

A frequently used method for studies of DNA ploidy of for example tumours is a detergent-trypsin method for the *preparation of cell nuclei* in combination with Ethidium or Propidium staining (Vindelov, Christensen & Nissen, 1983). Staining for DNA may be combined with protein stains of different fluorescent properties (Crissman, Oka & Steinkamp, 1976; Stoehr *et al.*, 1978). Optimal separation of the two fluorescence signals is achieved by sequentially exciting the cells in flow at two different wavelengths (Stoehr, Eipel & Goerttler, 1977; Dean & Pinkel, 1978). Recently, *in situ* hybridization methods for nucleic acids have been applied to cells in suspension for flow cytometric analysis (Trask *et al.*, 1985). With these methods specific base sequences can be demonstrated (Bauman, van der Ploeg & van Duijn, 1984).

Specific staining of RNA has been reported less frequently. Flow cytometric enumeration of reticulocytes on the basis of fluorescent RNA staining with Pyronin Y (Tanke *et al.*, 1980, 1983), Thioflavin T (Sage, O'Connell & Mercolino, 1983) and Thiazole Orange (Lee, Chen & Chiu, 1986) has been reported. Simultaneous staining of DNA and RNA using different dyes (Shapiro, 1981), or with the metachromatic dye Acridine Orange (resulting in green and red fluorescence of double- and single-stranded nucleic acids, respectively) (Traganos *et al.*, 1977) is feasible.

Immunofluorescent staining is described by Rost (1991b). Dual-laser flow cytometry enables flow cytometric studies of up to four colours simultaneously.

Fluorogenic enzyme substrates that can be applied to cells in suspension for flow cytometric analysis are very limited in number. One of the main reasons is the requirement for localization of the staining product at the enzymatic side, which is generally more difficult to fulfil for cells in suspension than on glass slides (Raap, 1986). Nevertheless, non-fluorescent staining of the enzymes peroxidase and esterase has proved to be useful for flow cytometric evaluation of differential blood cell counts as is done in the Technicon Hemalog D (Ansley & Ornstein, 1971). Fluorogenic substrates useful for flow cytometry are, among others, fluorescein diacetate (FDA) for esterases (Rotman & Papermaster, 1966), a number of peptide derivatives of naphthylamines for peptidases (Dolbeare & Smith, 1977), and naphthol AS-BI derivatives as substrates for phosphatases (Vaughan, Guilbault & Hackney, 1971). An overview of flow cytometric analysis of enzymatically stained cells has been given by Dolbeare & Smith (1979).

Cytophysical parameters

Signals measured by flow cytometry are generally due to the interaction of some form of energy and the biological material. In many cases energy is delivered as light rays, and analysis of cytophysical parameters may be based on the measurements of axial light loss (a rough measure for absorption), of fluorescence intensity (or related parameters such as fluorescence depolarization and anisotropy) and of light-scatter intensity (both in the forward direction and perpendicular to the incident light rays).

Electrical resistance (Coulter volume) measurements of biological objects in suspension were introduced by W.H. Coulter in 1949. His idea was to pass cells suspended in a conducting fluid through an orifice with a small diameter. Across the orifice, a constant current was maintained by two electrodes, positioned one on each side. Each cell that passed between the electrodes, being a relatively poor conductor due to its membrane composition, would change the electrical resistance and therefore, according to Ohm's law, generate a voltage pulse, which is easily amplified and measured (Fig. 14.3). As found by Coulter, this signal is proportional to the volume of the cells or particles. A volume distribution curve of objects thus can be derived from a recorded pulse height distribution. Coulter volume measurements obviously require differences in electrical resistance between the suspension medium and the objects. This can be easily achieved for live cells. It should be realized that cell death or fixation significantly reduces the electrical resistance of cellular membranes. Furthermore, cell volume measurements

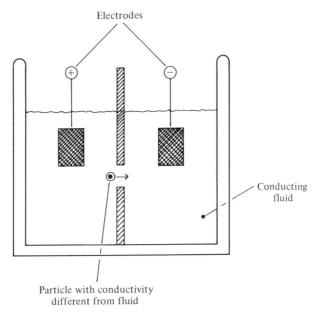

Fig. 14.3. Basic configuration for measuring electrical resistance according to Coulter. A constant current is maintained between two electrodes positioned in a conducting fluid. If cells with different conducting properties are passed through a small orifice, a change in electrical resistance occurs. This is observed as a voltage pulse (Ohm's law), which is directly proportional to the volume of the cells.

require an orifice size adapted to the size of the objects to achieve optimal resolution. Coulter volume measurements are widely applied for cell counting purposes in clinical haematology, e.g. counting of leukocytes, erythrocytes and platelets in blood samples. Furthermore, they are very useful in distinguishing different cell types in heterogeneous samples on the basis of size, and in distinguishing live and dead cells because of their difference in electrical resistance. An overview of electrical resistance pulse sizing has been given by Kachel (1979).

Light-scatter measurements are used in several flow systems for cell sizing and discrimination among cells with different morphologies (see also Fig. 14.4). Light-scatter signals from irregularly shaped biological objects are complex and sometimes difficult to interpret. If light, being an electromagnetic wave, is directed onto a cell immersed in a uniform medium, the electrons are slightly displaced relative to the nuclei of the atoms of the cell. This generates dipoles, which oscillate at the frequency of the incident light and reradiate light at the same wavelength as the incident light. The observed scattered light represents the summation of all reradiated waves, taking into account phase relationships. The signal depends on the size and refractive index distribution of the biological objects, the refractive index of the medium, the wavelength of the incident light and the angle of observation. For theoretical considerations the objects are assumed to be mostly spherical. Approximations for objects larger than the wavelength of the incident light have been made by Rayleigh–Debye. Application of this theory requires that the relative refractive index is close to 1 and that the phase

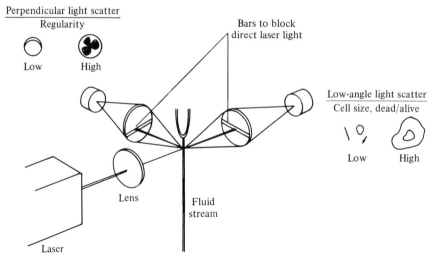

Fig. 14.4. Configuration for measuring forward and perpendicular light-scatter signals in an orthogonal flow cytometer. The fluid stream itself generates a horizontal plane of incident laser light, which is blocked by adjustable bars in front of each detector. Passage of an object through the laser beam will generate scatter signals at such angles, that they will pass the bars and are recorded by the detectors. Forward-angle light scatter is proportional to cell size and allows discrimination between live and dead cells. Perpendicular light scatter is influenced by the number of internal reflection sides of the cells, e.g. granularity and nuclear shape.

shift is small (Kerker, 1969). Narrow angle (forward) light-scatter signals for particles with a diameter d between roughly 5 and 10 μm are proportional to d^3. Beyond 10 μm, a relationship with d^2, and further on with d has been found (Mullaney & Dean, 1969).

Low-angle light scatter also allows discrimination between live and dead cells. This provides an important application, in particular for membrane immunofluorescence studies, where non-specifically stained dead cells may often be a problem. Light-scatter signals measured in a direction perpendicular to the incident light rays are strongly influenced by the regularity of internal cellular compartments. A cell with an irregularly shaped nucleus and/or cytoplasmic granules will generally have a higher perpendicular light scatter than an agranular cell with a round nucleus. Measurements of forward and perpendicular light-scatter signals can be used to distinguish different types of cells in blood and bone marrow (Salzman, Growell & Martin, 1975). Light-scatter signals are predominantly measured using orthogonal laser-based systems. However, microscope-based flow cytometers have been adapted for scatter measurements as well (Steen, 1980).

In comparison to Coulter volume measurements, forward light scatter provides comparable information. Forward angle scatter measurements allow the detection of particles as small as approximately 0·2 μm depending on their refractive index, with standard flow cytometric equipment. The lower detection limit for Coulter volume measurements is in the order of a few micrometres for most systems. Moreover, light-scatter signals do not require a special sensing orifice as is necessary for Coulter volume measurements, since in many cases a focussed laser light beam is used to generate fluorescence.

Fluorescence is a very complex luminescence phenomenon. A large number of independent fluorescence parameters have been studied in spectroscopy, such as total intensity, excited state lifetime, excitation and emission spectra, emission anisotropy, fluorescence energy transfer and several others (Dr Tomas Hirschfeld (personal communication) once listed more than 30 fluorescence-related parameters that reveal information about the fluorophore and the biological structure that contain it). Although many of these parameters can be studied in principle by flow cytometry also, the number of fluorescence parameters generally studied in flow has been limited so far. The majority of studies concerns the measurements of total fluorescence intensity. In addition, *fluorescence polarization* and *energy transfer* studies are regularly performed.

The degree of polarization of emission depends on the motility of the fluorophore, and therefore gives information on its vicinity. Special dyes, such as diphenylhexatriene, may be incorporated into the cell membrane. Measurements of the degree of polarization of such dyes allow the study of membrane fluidity. Energy transfer between two fluorophores may occur if the absorption and emission spectra of the two are (partially) overlapping: one fluorophore absorbs the excitation energy, which is transferred to the other and used as excitation energy to generate emission. Energy transfer strongly depends on the distance (s) between the fluorophores (proportional to s^6), and therefore occurs only when the separation is less than 10 nm. It therefore is a useful parameter for the study of proximity relationships. Extensive discussion of fluorescence polarization and energy transfer is not feasible here. A detailed review of theory and flow cytometric applications has been given by Jovin (1979).

The intensity of the fluorescence emission, as derived from measured peak height or area, offers an extremely versatile parameter. In the first place, there is available a variety of fluorescent dyes that specifically bind to cellular macromolecules, based on defined chemical or physical interactions. Examples are fluorescent basic dyes, that contain primary amino groups, by which they can be bound to aldehyde groups (Schiff reaction). Such aldehyde groups may be generated by oxidation of sugars as occur in polysaccharides and glycoproteins, or by acid hydrolysis of DNA. Moreover, fluorescent dyes may change their emission properties as a function of pH or ionic composition (see Chapter 5). Using such dyes, flow cytometric determinations of intracellular pH or ionic calcium concentration have been described. An important group of fluorophores consists of the dyes that are coupled to antibodies for immunofluorescence studies.

The major advantage of the use of fluorophores for flow cytometry is their *sensitivity*. Modern flow cytometers allow the specific detection of a few thousand fluorophores per cell, and are therefore very suitable for demonstrating or quantifying low amounts of cellular macromolecules. This has been achieved by combining high-power lasers, efficient light collecting optics and sensitive photodetectors. The lower detection limit of fluorescence measurements in flow is therefore determined mainly by the level of *autofluorescence* of the biological object, which may be several thousands of fluorescein equivalents (Aubin, 1979; Jongkind *et al.*, 1982), and autofluorescence of the lenses and filters when high excitation intensities are used. It is possible to correct for autofluorescence using dual-laser differential fluorescence measurements (Steinkamp & Stewart, 1986). Another possibility is the use of luminescent dyes that show delayed fluorescence, such as chelates of the lanthanides, or phosphorescence. These dyes allow time-resolved measurements of the delayed signal (excited state lifetime ranging from micro- to milliseconds), thereby eliminating the disturbing autofluorescence (lifetime generally less than a nanosecond). Applications of such measuring techniques to dyes in cuvette photometers have been shown to be at least one order of magnitude more sensitive than conventional fluorometric procedures (Soini & Kojala, 1983). Time-resolved luminescence measurements of cytochemically stained cells in suspension may further increase the sensitivity of flow cytometry, and thereby increase its applications.

Data presentation and analysis

Flow cytometers are able to measure simultaneously several parameters from one object. A typical number for clinically used instruments is 3–5; however, larger experimental systems allow measurements of up to 8 parameters. After detection and amplification, analogue signals are first digitized using analogue-to-digital (AD) converters. AD conversion in flow cytometry must be fast. For example, to record five-parametric data at a speed up to 2000 cells/s, digitizing and storing of a single signal must be performed within 20 μs. At higher rates, as used in high-speed cell sorting, the complexity of data acquisition increases.

Digitized data may be displayed directly using univariate or bivariate frequency distributions (histograms). Examples are given in Fig. 14.5. The display of more than two parameters (multivariate data) becomes very complex. A third parameter may be displayed in the vertical direction. The result is the display of a cube containing dots, each point representing the location of a parameter triplet of a corresponding cell in this cube. Another possibility to simplify visualization of multiparameter flow data is to use colour for the third parameter, and colour intensity for frequency (Stoehr & Futterman, 1979). Multiparametric data are preferably recorded in list mode. This means that the correlated data are digitized and stored directly on disc or tape, while the sample is being analysed. Selection of the most discriminative parameters as well as gated analysis of the data may then be performed later. Flow cytometric data may contain a wealth of biologically relevant information. Mathematical analysis of the data is often required to extract this information. Most analysis is performed on univariate data, although there is an increasing interest in quantitative analysis of multivariate data (partly explained by the dramatically improved computing facilities for this purpose). Methods in use range from graphic procedures through non-linear curve-fitting procedures to cluster analysis. A recent overview of this subject has been given by Dean (1985).

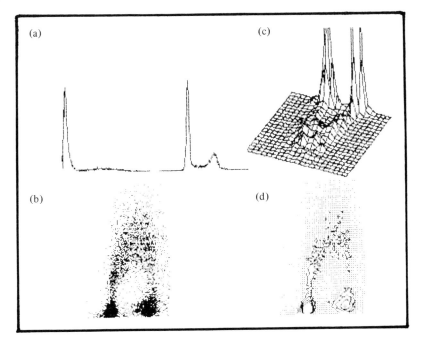

Fig. 14.5. Examples of univariate and bivariate histograms for simultaneous analysis of DNA content (red Propidium fluorescence) and incorporated BrdU (green FITC anti-BrdU fluorescence) in cultured breast tumour cells in S-phase at the moment of incubation with BrdU. (a) Single parameter distributions of DNA content and FITC anti-BrdU fluorescence, respectively. (b) Correlated two parameter data (dot plot). (c) Pseudo-3-dimensional display of correlated two-parameter data. (d) Contour map: projection of the data shown in (c) with contour lines set at defined numbers of cells.

Clinical applications of flow cytometry

Clinical applications of flow cytometry are numerous and rapidly increasing. A complete overview is therefore hardly possible. Typical examples of clinical applications of flow cytometry have therefore been selected.

The earliest clinical applications of flow cytometry were in the field of haematology, where Coulter counters were introduced for the counting of blood cells such as erythrocytes, leukocytes and platelets. Later on, measurements of light scatter and axial light loss were performed for differential counting of leukocytes based on differences in peroxidase and esterase staining. At present, flow cytometric analysis is increasingly being used in haematology for the counting of reticulocytes in peripheral blood and for the immunological characterization of leukaemias.

The number of reticulocytes (normal value around 2%) in the peripheral blood reflects the erythropoietic activity of the bone marrow, which is of considerable interest in case of anaemia and in oncological patients treated with cytostatic drugs that affect bone marrow proliferation. However, accurate counting of especially low numbers of reticulocytes (1% or lower) form a major problem in routine haematology, due to the large statistical errors involved in determining such percentages.

Flow cytometric analysis of reticulocytes, fluorescently stained for RNA with Pyronin Y (Tanke *et al.*, 1980), has proved to be of considerable value. Within one minute, 100000 cells may be readily analysed and the results plotted as RNA histograms (Fig. 14.6). The percentage of reticulocytes as well as the age distribution (younger reticulocytes contain more RNA than do older ones) can be calculated from these profiles (Tanke *et al.*, 1983). Besides Pyronin Y, other dyes such as Acridine

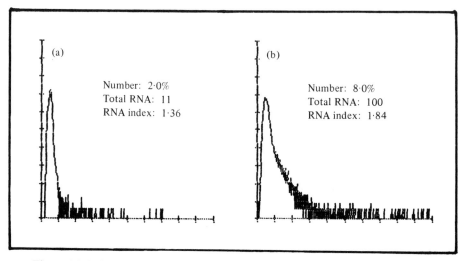

Figure 14.6 Fluorescence histograms obtained by flow cytometry of Pyronin Y stained red blood cells. Relative Pyronin Y fluorescence is proportional to RNA content of the reticulocyte. (a) Normal blood sample with 2% reticulocytes; (b) blood sample with 8% reticulocytes. Note the presence of immature reticulocytes with increased RNA content in the higher fluorescence channels.

Orange (Natale, 1982), Thioflavine T (Sage *et al.*, 1983) and Thiazole Orange (Lee *et al.*, 1986) have been proposed. Flow cytometric analysis of reticulocytes is more reproducible than manual evaluation (coefficient of variation approximately 20% versus errors up to 100% as reported for hand counts), and is sensitive enough to detect changes as a consequence of chemotherapy (Tanke *et al.*, 1986). Current efforts in this field concentrate on the development of appropriate standards, such as microspheres with defined amounts of RNA, that can be mixed with blood and used as internal standards to define the reticulocyte population. The use of such an arbitrary definition of a reticulocyte has the advantage that the outcome of the test becomes more or less independent of the type of flow cytometer and fluorescent RNA-staining dye that is used. At present, normal values have to be determined for each staining procedure, type of instrument and mathematical analysis procedure.

The classification of leukaemias has been based on morphology of the cells and their cytochemical staining properties. Recently various monoclonal antibodies have become available that recognize certain epitopes on the cell membrane of the leukaemic cells. Especially when combinations of different reagents are used, leukaemias can be immunologically characterized (Foon, Schroff & Gale, 1982; Majidic *et al.*, 1984). Multiparametric studies are difficult and time-consuming when performed manually. Flow cytometry offers an excellent tool to perform these tests quickly and reproducibly.

The analysis of DNA ploidy in clinical pathology is another major application of flow cytometry. Cells with abnormal DNA content are often observed in neoplasms. Provided that a sufficient number of these abnormal cells is present, DNA flow histograms may yield diagnostic information. Especially in those cases where clinical examinations and conventional morphological and cytochemical methods do not give a decisive answer, observed DNA abnormality strongly indicates a neoplastic process. Besides diagnostic values, DNA ploidy of the tumour has been evaluated as a prognostic parameter. Generally, tumours with diploid or peri-diploid stemlines (see Fig. 14.7) have a better prognosis in comparison with tumours with stemlines of aneuploid high DNA content, although this may depend on the type of tumour and the applied therapy. A major disadvantage of prognostic studies in oncology is the long time of follow-up (sometimes up to 10 years) that is required before the real clinical value of such prognostic parameters can be statistically verified. The use of paraffin-embedded archival material for flow cytometric analysis of DNA content, originally described by Hedley *et al.* (1983) allows retrospective analysis of a well-documented material with known follow-up. Moreover, it has also been possible to study the expression of onco-proteins in paraffin-embedded material (Watson, Sikora & Evan, 1985), although their diagnostic and prognostic importance for the various tumours is not clear yet.

Further exploration in the field of DNA is to be expected when *in situ* hybridization methods can be applied to localize specific sequences of nucleic acids in interphase cells and chromosomes, using non-autoradiographic methods (Bauman *et al.*, 1984). Rapid analysis of large numbers of cells or chromosomes, as well as quantification of the hybridization reaction as a parameter for gene copy number and/or expression can be performed by flow cytometry, provided that the preparative problems of *in situ* hybridization reactions in suspension can be solved (Trask *et al.*, 1985).

Transplantation immunology makes use of flow cytometry for the measurement of subpopulations of mononuclear cells, in particular T cells. The steadily increasing number of monoclonal antibodies by which mononuclear blood cells, with somewhat different functions, can be recognized, allows characterization of the immune system in more detail and more easily than before. Although studies of lymphocyte subsets can also be done using microscopic techniques, flow cytometry is at present the method of choice. Manual counting of a few hundred cells is not only time consuming and therefore expensive, but also leads to large statistical errors. Both can be overcome by rapid, objective flow cytometric analysis of 10 000 cells in a few seconds. Flow cytometric evaluation becomes almost a prerequisite if subsets have to be identified on the basis of three- or four-colour immunofluorescence (Parks, Hardy & Herzenberg, 1984).

The so-called T4/T8 ratio (roughly related to T helper cells/ T suppressor cells) has been extensively studied in renal and bone marrow transplantation (BMT). Pre-transplant T4/T8 ratios were found to correlate with renal graft survival in adult patients on low-dose corticosteroid therapy: low ratios correlated significantly with poor graft survival (van Es *et al.*, 1983). This was, however, not observed in patients treated with cyclosporin A. The study of T4 and T8 cells in patients treated with methotrexate for BMT showed a lower repopulation rate for T8 cells in comparison with T4 cells. However, patients that developed at least Grade 2 graft versus host disease (GVHD) showed a much faster repopulation rate of the T8 subpopulation, in comparison with those patients that developed no or only Grade 1 GVHD. Measurement of T4/T8 ratio at day 12 after transplantation therefore allowed the identification

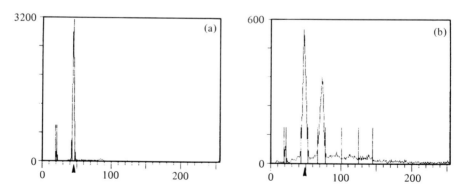

Fig. 14.7. Fluorescence histograms recorded by flow cytometry of Ethidium-stained human tissue (according to Vindelov *et al.*, 1983). (a) Normal diploid DNA distribution. (b) Abnormal aneuploid DNA distribution obtained from a bladder tumour. Besides a normal cell population with diploid DNA content (DNA index = 1·0) an aneuploid tumour population (DNA index = 1·7) is observed. Arrows indicate diploid DNA content based on measurements of internal reference cells (chicken erythrocytes) with known DNA content.

of patients at risk for severe GVHD (Gratama *et al.*, 1984). As in renal transplantation, this phenomenon was not observed in patients treated with cyclosporin A.

Recently developed monoclonal antibodies allow the specific recognition of the various types of cytotoxic lymphocytes: natural killer cells (non-T cells), HLA class 1 restricted cytotoxic T cells, and HLA non-restricted T cells (Lanier & Philips, 1986). It is to be expected that with this type of reagent, more refined information about the immune status of transplant patients can be obtained.

Future prospects of flow cytometry

In the past decade, two developments have significantly influenced the applicability of flow cytometry for both experimental studies and clinical applications. First, the hybridoma technique, by which mono-specific antibodies can be prepared in almost unlimited amounts, has facilitated the quantitative studies of a variety of macromolecules and cell biological processes that were hardly possible before that time. Secondly, the use of microcomputers for operation of flow cytometers and for adequate data handling and presentation has been clearly established. Their application has changed flow cytometers into user-friendly instruments that can be operated easily even by relatively inexperienced personnel. This aspect has considerably stimulated the use of flow cytometers in clinical laboratories.

Further use of monoclonal antibodies and specific fluorescent dyes, as well as further perfection of the instrumentation and lowering the costs of operation, will enlarge the contribution of flow cytometry in experimental and clinical studies. A number of clinical tests such as counting of reticulocytes and platelets, phenotyping of lymphocytes and immunological characterization of leukaemia, which are still performed manually in many clinical centres, are expected to be carried out using flow cytometry.

It should be noted that also in the field of image analysis (Chapter 15), significant progress along these lines has been made recently. Image analysis has and will profit in a similar way from the availability of a variety of immunological reagents and powerful microcomputers, and is now fast enough to analyse a few hundred cells/second. For a number of clinical applications image analysis may therefore compete with flow cytometry in the near future. However, a major advantage of flow cytometry is its flexibility in sorting live cells and cellular constituents, which will ensure its unique position in biology and medicine.

15

Scanning, video intensification and image processing

This chapter deals with a variety of techniques in which the fluorescent specimen is scanned, measurements of the fluorescence intensity being taken during the scanning, enabling an image to be built up as a video frame. The scanning may be carried out mechanically or with a television camera. The final image is commonly held in digital form, and can be subjected to various forms of image processing by computer.

Why scan? We have already seen that, for quantification, the use of fluorescence obviates the need for scanning which, in absorptiometry, is required to avoid distribution error. However, scanning of a fluorescence image may be useful for other reasons, where morphological features and fluorescence properties can usefully be correlated. For quantification of catecholamines in nerve terminals, scanning methods are general seen as the methods of choice. Scanning of the image with a television camera, with the televised image shown on a video monitor, enables an otherwise dim image to be seen more conveniently. Time-lapse video recording can be used to study movements of fluorescent probes in living cells. Digitization of the image permits image analysis, including quantification, exactly as in absorptiometry. The term fluorescence digital imaging microscopy (F-DIM) has been applied to this technique (Arndt-Jovin *et al.*, 1985). If both the illumination and detection systems are simultaneously focussed onto a spot which scans the specimen, this is called confocal fluorescence microscopy, and has several advantages which will be described below. Scanning and some form of image analysis is increasingly being applied to the routine screening of smears for the detection of cancer.

For comparison, the system in a conventional (non-scanning) microscope is illustrated in Fig. 15.1. In a scanning fluorescence microscope, the specimen can be

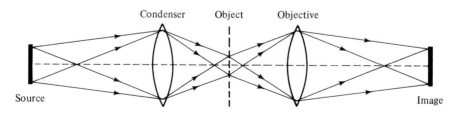

Fig. 15.1. Schematic diagram of a conventional fluorescence microscope.

162

scanned in any of three ways, as follows:

Type Ia. With the specimen uniformly illuminated, the emission is scanned. Scanning may be made either by a video camera (Fig. 15.2a), or by mechanical means whereby a pinhole aperture is placed in front of a photomultiplier tube and either the specimen is moved in raster fashion or mirrors shift the image of the object in corresponding fashion across the pinhole.

Type Ib. With the illumination confined to a focussed spot, preferably produced with a laser, and emitted light measured from the entire field (or at least an area large in relation to the excited spot), either the specimen is scanned mechanically in raster fashion with respect to the excitation spot, or the spot is moved ('flying spot') so as to scan the specimen (Fig. 15.2b).

Type II. Both the area of illumination and the area from which measurement is made are confined to a spot, the laser (or other point source) and a pinhole aperture in the emission-measuring system being focussed on the same spot in the object plane; either the spot or the specimen is moved mechanically in raster fashion (Fig. 15.3). Because both illumination and detection systems are confined to points focussed on the same spot, this is known as *confocal microscopy.*

During the process of scanning, emission is measured by a detector, which may be a photomultiplier tube, solid-state device, or video camera. The output may be recorded either as an analogue signal or digitally. In the latter case, each number recorded

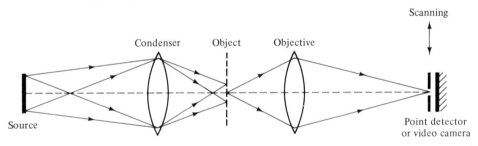

Scanning Type Ia

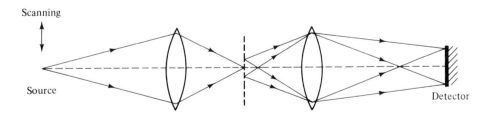

Scanning Type Ib

Fig. 15.2. Schematic diagram of scanning Type Ia and b. For explanation, see text.

represents the emission intensity in respect of a small region of the specimen known as a *pixel*. The dimensions of the pixels determine the resolution of the picture, and the total number of pixels in the scanned area determines the amount of detail in the final image. Since the spectral sensitivity of scanning fluorescence microscopy (SFM) is limited only by the detector, studies are possible concerning otherwise unvisualizable fluorescence in the ultraviolet (UV) and infrared (IR). In the IR, SFM has been applied to Indocyanine Green (Moneta *et al.*, 1987), and should be ideal for studying chlorophylls. There are numerous substances which are fluorescent in the UV, hitherto studied biochemically and which could perhaps now be studied by fluorescence microscopy.

Scanning has in principle a number of advantages. Firstly, the picture produced on a monitor screen is much brighter than the view in the microscope, obviating the necessity for a darkened room and dark-adapted eyes, and the greater sensitivity allows reduction of the intensity of irradiation, thereby reducing fading and damage to living tissues. Secondly, the video screen can easily be viewed simultaneously by two or more observers. Thirdly, the video output can be recorded either continuously or intermittently (time-lapse), for future study. Fourthly, taking any frame, or an average of several frames, a computer can analyse the data, present a modified version (e.g. with increased contrast) on the monitor screen, calculate areas or integrated intensity over a specified area, and transmit the picture directly to a graphics printer to produce a picture on paper. Alternatively, the monitor screen can be photographed; the exposure required is much shorter than direct photography of the fluorescence, and the exposure time is constant. Finally, use of a photomultiplier, video camera or charge-coupled device (CCD) expands the usable emission wavelengths to beyond the visible, to the UV and IR. Scanning with an excitation beam has the additional advantage that only one minute area (that of the scanning spot) is irradiated at any one time, and then only

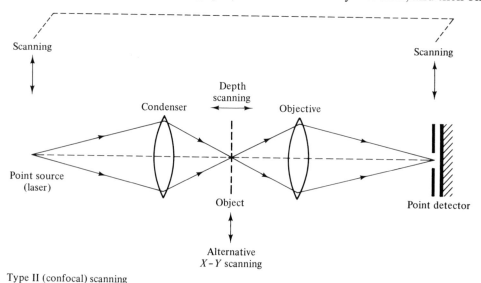

Fig. 15.3. Schematic diagram of scanning Type II (confocal).

briefly, so that photobleaching of the specimen is reduced. Moreover, there is only minimal generation of autofluorescence in the optical system, thereby increasing the signal/noise ratio and the overall sensitivity (Ploem, 1986). The main disadvantages of scanning are complexity and expense of the apparatus, and possibly limited detail in the image (determined by the pixel size).

Scanning optical microscopy, fluorescence and conventional, was reviewed by Wilson & Sheppard (1984), who deal with the mathematical basis in some detail. See also Ash (1980).

TYPE IA SCANNING:
VIDEO FLUORESCENCE MICROSCOPY

Scanning the emission while exciting the complete field is most easily accomplished by examining the image with some form of closed-circuit television (CCTV) system. This technique is now generally known as video intensification microscopy (VIM), so named by Willingham & Pastan (1978). However, since video intensification can in principle be applied to other modalities of microscopy besides fluorescence, perhaps it might be generally more appropriate to refer to video fluorescence microscopy (VFM). Video microscopy in general is described by Inoué (1986), who also gives a useful glossary of technical terms; see also Bright & Taylor (1986), Arndt-Jovin *et al.* (1985), Inoué (1988), Jarvis (1988) and Shotton (1988a).

Instrumentation

Video fluorescence microscopes
Instrumentation for video fluorescence microscopy has been reviewed by Arndt-Jovin *et al.* (1985) and described by several authors, including Willingham & Pastan (1978), Bollinger *et al.* (1979), Bollinger, Franzeck & Jäger (1983), Rich & Wampler (1981), Franzeck *et al.* (1983b), Yanigida *et al.* (1983), Berns, Siemens & Walter (1984), Benson *et al.* (1985), Williams *et al.* (1985), Feuerstein & Kush (1986) and Webb & Gross (1986). Video microscopy in general was described in detail by Inoué (1986). A system for simultaneous Nomarski and fluorescence imaging for video microscopy was described by Foskett (1988).

Video cameras
In a conventional television camera, an optical image is projected onto a photocathode and scanned by an electron beam. The output of the camera is an electrical signal proportional to the intensity of the optical image at the point being scanned at the moment. The picture is scanned in horizontal lines, either 625 or 525 lines in the full height of the picture. Alternate lines are scanned in successive frames, so that the complete image is scanned over two successive frames. Frames are obtained either 50 or 60 times per second, depending on the frequency of the local electricity supply (50 Hz, except in the United States and some other countries which use 60 Hz). The output from the camera includes also synchronization pulses to keep the receiver in synchrony.

To date, only black-and-white video has been used. Unfortunately, ultrasensitive colour video cameras are still very expensive; in the next edition I hope to report on video microscopy in full colour! Television technology for microscopy is described in detail by Inoué (1986); see also Jarvis (1988).

Charge coupled devices

A charge-coupled device (CCD) is a solid-state photometric device on a semi-conductor chip. The light-sensitive region, currently (1990) typically of the order of 8 mm × 6 mm, is divided into rows and columns of smaller areas (pixels), typically 512 × 512 pixels or 600 × 500 pixels. Photons striking the pixels cause an electric charge to accumulate at a rate proportional to the intensity of the light. At regular intervals, the amount of the charge on each pixel is read out, a row at a time, as a video frame or to a computer. For details of these devices, see Eccles, Sim & Tritton (1983), Hobson (1978) and Mobberley (1987).

Recently, video cameras have been developed using CCD sensors. CCD cameras have several advantages over photomultiplier tubes for cytophotometry, including inherent linearity, a relatively flat spectral response in the visible region, geometric stability, no image persistence and resistance to damage by excessive illumination; on the other hand, the signal-to-noise ratio tends to be low. The relative merits of CCD and conventional video cameras have been discussed by Spring & Smith (1987) and Jarvis (1988). A cytophotometer for absorption measurements with a CCD video camera has been described (Donovan & Goldstein, 1985).

Image intensification

An image intensifier is an electronic device which accepts a dim image and produces a brighter image on a fluorescent screen. Its electronic amplification is similar to that of a photomultiplier tube. One important type of image intensifier was very clearly described by Lampton (1981). For black-and-white images only, an image intensifier can provide initial amplification of the image for a television or CCD camera. Gawlitta et al. (1980) employed a Siemens compact K5B camera with an integral image intensifier (Siemens) to follow pinocytosis and locomotion of amoebae with chlorotetracycline fluorescence. They found three advantages in using this system: the living cells were not subjected to the damaging influence of intensive irradiation, the fluorescent probe was not bleached during irradiation, and the rapid dynamics of Ca^{2+} fluxes could be recorded using short photographic exposure times (250–500 ms). See also Blanchard, Chang & Cuatrecasas (1983), and Spring & Smith (1987).

Photography from monitor screen

Data from scanning microscopy is commonly recorded either in analogue form as videotape, or in digital computerized form. In addition, images from the monitor screen (video or computer) can easily be photographed. My own practice is to set up a camera on a tripod on the floor in front of the screen. Using a 35 mm single lens reflex (SLR) camera, a telephoto lens of about 100 mm to 135 mm focal length is convenient as it allows the camera and its tripod to be sufficiently far away from the screen to leave the area in front clear, and also to avoid distortion due to photographing a curved screen from too close. Screens may be designed to be viewed from a certain distance; in such case the camera lens should be at that distance from the screen. If a SLR camera is not available, particular care will need to be taken to compose the picture in the viewfinder, allowing for parallax (due to the lens and the viewfinder being in different places).

Almost any black-and-white or colour film of normal contrast can be used. The shutter speed should be long enough to cover several frames of video, otherwise the raster pattern will appear. This problem is particularly apparent with cameras with focal plane shutters (as in most SLR cameras), which record a diagonal stripe if a short exposure is used. Unless the subject is moving rapidly, it is well to use a shutter speed of about 1/8 to 1/2 s. The brightness setting of the monitor should be noted; if necessary, the brightness control knob should be calibrated (say 0–10, or 0–8 which is easier to determine by eye). Normally the brightness setting will not be changed, once set to a suitable level, and the exposure required for the camera will also remain constant. It is preferable not to use the 'auto' setting of an automatic camera, since this may inappropriately vary the exposure from one experiment to the next. Because the screen is relatively flat, depth of field is not likely to be a problem, so that any convenient lens aperture can be used.

Quantification

Quantitative video intensification microscopy (QVIM) is a logical extension of VIM. Benson, Knopp & Longmuir (1980) reported that their instrument could measure fluorescence intensities with an error of 1·5% of full scale in 65 536 different positions in a microscope field; a video frame freeze acquisition time of 33 ms allowed time-dependent changes of that order of time or slower to be followed.

Applications

Qualitative applications

The major application of VIM has been to dynamic studies of living cells, where the particular advantages of VIM are most felt. These applications include studies of uptake of fluorescently labelled proteins by cultured cells (Willingham & Pastan, 1978; Maxfield et al., 1978; Schlessinger et al., 1978) binding of a fluorescent analogue of enkephalin to neuroblastoma cells (Hazum, Chang & Cuatrecasas, 1979); distribution of fluorescently labelled actin and tropomyosin after microinjection into cultured cells (Wehland & Weber, 1980); intracellular calcium ionic concentration using chlorotetra-cycline as indicator (Gawlitta et al., 1980); motility of human spermatozoa labelled with fluorescein isothiocyanate (FITC) or tetramethylrhodamine isothiocyanate (TRITC) (Blazak et al., 1982); capillary angiography in humans using Uranin (sodium fluoresceinate) (Bollinger et al., 1982, 1983; Bollinger, Franzeck & Jäger, 1983; Franzeck et al., 1983a,b; Baer et al., 1985; Bollinger, 1988) or Indocyanine Green

(Moneta et al., 1987); DNA molecules in solution (Yanigida et al., 1983, 1986); mitochondrial patterns in living myocardial cells in vitro, stained with Rhodamine 6G (Berns et al., 1984); distribution and movement of endosomes, lysosomes and mitochondria in cultured rat ovarian granulosa cells using pyrene-concanavalin A (P-Con A) and 3,3'-dioctadecylindocarbocyanine-labelled low-density lipoprotein (diI-LDL) (Herman & Albertini, 1984); lateral transport in membranes (Kapitza, McGregor & Jacobson, 1985); uptake of benzo[a]pyrene by living cells in culture (Plant, Benson & Smith, 1985); blood platelet adhesion using Quinacrine-labelled platelets (Feuerstein & Kush, 1986); various probes in living cells, particularly studying actin (Taylor et al., 1986); hepatic transport of sodium fluoresceinate and FITC-labelled sodium glycocholate (Sherman & Fisher, 1986); molecular motions of receptors on cell membranes for low-density lipoprotein (Webb & Gross, 1986); and cell-to-cell diffusion of fluorescent tracers (Safranyos et al., 1987).

For static studies, Forman & Turriff (1981) found VIM convenient, noting that 'the advantages of VIM are greatest in applications where convenience in handling and recording images from a large number of specimens is more important than achieving the best possible image quality'. See also Vrolijk et al. (1980). Video techniques can also be applied to surface fluorescence by using a macrophotographic system (Daley et al., 1989). In the future, VIM is likely to find valuable application for studying fluorescence of substances which fluoresce in the IR (such as chlorophylls) or the UV (such as tryptophan).

Quantitative applications

The methodology for QVIM has been investigated by Sisken, Barrows & Grasch (1983, 1986) and Sisken et al. (1985). Benson et al. (1980) used QVIM for intracellular oxygen measurements. Barrows et al. (1984) measured nuclear DNA using bisbenzimide. QVIM is useful for studies of fluorescence recovery after photobleaching (FRAP; see Chapter 11); see Benson et al. (1985) and Koppel, Carlson & Smilowitz (1989).

TYPE IB SCANNING: SCANNING WITH THE EXCITATION BEAM

An excitation-scanning fluorescence microscope is rather similar in principle to a scanning electron microscope; the specimen is scanned at the excitation wavelength by a narrow beam of light from a laser, and the emitted light is detected, quantified, and the resulting pixels built up into an image by a computer. It has already been indicated that use of a narrow excitation beam irradiating only one minute area at any one time, and then only briefly, reduces photobleaching of the specimen and the generation of autofluorescence in the optical system, thereby increasing the signal-to-noise ratio and the overall sensitivity.

Scanning of fluorescence was introduced by Mellors & Silver (1951), Mansberg & Kusnetz (1966), and Gillis et al. (1966), who used event-counting systems. The original apparatus of Mellors & Silver (1951) was primitive by present-day standards. The specimen was scanned by a 'flying-spot' for excitation, generated by a Nipkow disc (a spiral series of apertures in a revolving disc) in the excitation pathway (Fig. 15.4). Output from a photomultiplier was fed to a cathode-ray tube and to an event counter whose purpose was to count the number of cells with fluorescence above a present level. Application to cancer diagnosis was described by Mellors, Glassman & Papanicolaou (1952). Modern systems employ a laser beam, scanned in a raster over an area of the specimen using oscillating mirrors driven electrically (Slomba *et al.*, 1972; Shack *et al.*, 1987).

Applications

The concept of automated scanning in a fluorescence microscope of a diagnostic specimen for detecting neoplastic (cancer) cells appears to have been originated by Mellors (Mellors & Silver, 1951; Mellors *et al.*, 1952). More modern systems are those of Golden *et al.* (1979), who utilized the Leitz 'Texture Analyzer System', and Parry & Hemstreet (1988). I believe that commercial systems are now available or are projected for the near future.

Fluorescence recovery after photobleaching

Photobleaching in the FRAP technique (see Chapter 11) is usually carried out with a laser. The laser beam is commonly confined to a single spot or fixed pattern, but can be moved to provide multipoint data (Koppel, 1979) or scanning (Wade, Trosko & Schindler, 1986).

Catecholamines

The microfluorometric quantification of catecholamines (and serotonin) has been referred to in Chapter 5. A major difficulty with the histochemical quantification of

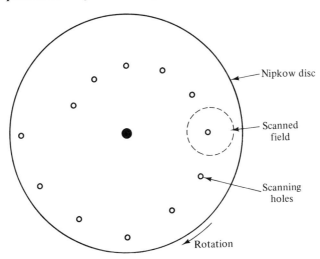

Fig. 15.4. A Nipkow scanning disc, similar to that used by Mellors & Silver (1951) for excitation scanning. As the disc revolves, each aperture in turn passes across the field; successive apertures in the spiral scan a series of lines across the specimen.

catecholamines in nerve fibres is that the catecholamines are irregularly distributed; the catecholamines tend to be concentrated in varicosities along the length of nerve fibres, and the fibres themselves are irregularly distributed in the tissue. Rather than measurements from arbitrarily defined circular or rectangular areas, including both nerves and background tissue, it has been found preferable to scan large areas and to use image analysis to determine the mean catecholamine concentration in the fluorescent areas assumed to correspond to the actual nerves (Schipper, Tilders & Ploem, 1978, 1979, 1980; Schipper & Tilders, 1979; Schipper et al., 1980; Alho, Takala & Hervonen, 1980; Cowen & Burnstock, 1982; Henschen & Olson, 1983; Larsson, Goldstein & Dahlström, 1984).

TYPE II SCANNING:
CONFOCAL FLUORESCENCE MICROSCOPY

We have already seen that, in a confocal fluorescence microscope, both the area of illumination and the area from which measurement is made are confined to a single spot; the laser (or other point source) and a pinhole in the emission-measuring system are focussed on the same spot in the object plane, and either the specimen or the spot is moved in raster fashion. While substantially complicating the microscope, this arrangement leads to several significant advantages, particularly for fluorescence microscopy. Firstly, a confocal microscope possesses superior imaging capabilities compared to a normal microscope (Sheppard & Choudhury, 1977); for a given numerical aperture, the lateral resolution of the system can be greater by a factor of 1·4 (Brakenhoff, Blom & Barends, 1979), and further improvement in both lateral and axial resolution is possible (Bertero et al., 1990). Moreover, in fluorescence microscopy, the resolution of the scanning system is dependent on the excitation wavelength, rather than the emission wavelength as in a conventional fluorescence microscope, and therefore greater resolution is obtainable (Cox, Sheppard & Wilson, 1982). Secondly, the confocal microscope possesses a unique depth-discriminating property (Sheppard & Wilson, 1978b; Hamilton, Wilson & Sheppard, 1981), whereby the detector does not accept all the light from out-of-focus planes and so these are imaged less strongly than the in-focus plane. In consequence, background fluorescence may be sharply reduced (Cox, 1984), and it is possible to obtain three-dimensional data by stereoscopic imaging (Brakenhoff, 1979) or by building up a series of optical sections (Wijnaendts van Resandt et al., 1985; Shotton, 1988b; Takamatsu & Fujita, 1988). For optical sectioning, best results are obtained if the wavelengths used for excitation and for emission measurement are close to one another; the longer the emission wavelength, the poorer the depth discrimination (Wilson, 1989). Fluorescence microscopy with epi-illumination lends itself particularly well to confocal scanning, since the illumination (excitation) and observation (emission) optical pathways overlap.

A schematic diagram of a generalized confocal microscope is shown in Fig. 15.3, and that of a confocal fluorescence microscope in more detail in Fig. 15.5. The basic design of confocal microscopes is due to Minsky (1957); for a detailed review see Wilson & Sheppard (1984). There are two main types of confocal microscope. In one type, known as a tandem scanning reflected light microscope, the scanning is performed rapidly with multiple beams, and the image is viewed directly. In the other type, usually known as a laser scanning confocal microscope, or unitary beam confocal scanning microscope, scanning is performed with a single beam, fluorescence is measured and the image is built up as a video frame.

The tandem scanning confocal fluorescence microscope (TSRLM; Petráň *et al.*, 1968, 1985; Petráň, Hadravsky & Boyde, 1985) has illumination in the form of multiple narrow beams from an array of apertures focussed by the objective into the focussed-on plane of the specimen. Fluorescence from the illuminated patches of the specimen is imaged onto an identical array of apertures in the intermediate image plane of the microscope. Both arrays of apertures are scanned in tandem – hence the name tandem scanning. All extraneous light from out-of-focus layers or scattered from optical surfaces in the microscope is intercepted by the solid portions of the aperture array in the intermediate image plane or by light-traps in the microscope. The TSRLM is used like a conventional light microscope: the observer looks through the eyepiece at a steady image. Focussing up and down, a mental image can be constructed of three-dimensional structures in the usual way. Further details of the fluorescence mode of the TSRLM are given by Boyde (1985, 1987), Boyde & Watson (1989) and Boyde *et al.* (1990).

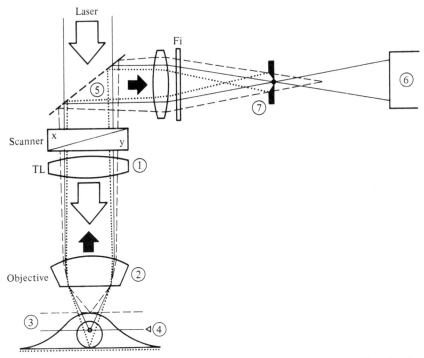

Fig. 15.5. Diagram of a confocal fluorescence microscope (Zeiss LSM). Key: 1, tube lens; 2, objective; 3, specimen; 4, focal plane; 5, beamsplitter; 6, detector; 7, pinhole aperture; TL, tube lens; Fi, filter. Out-of-focus contributions are efficiently blocked by the spatial filter (7). Diagram courtesy of Carl Zeiss Pty Ltd, Sydney.

Unitary-beam confocal fluorescence microscopes scan in one or other of three ways: (1) simultaneous movement of images of the illumination and observation apertures over the specimen, using oscillating mirrors (Wilke, 1985; White, Amos & Fordham, 1987; Carlsson, 1990) or acousto-optical devices (Horikawa, Yamamoto & Dosaka, 1987; Draaijer & Houpt, 1988); (2) the illumination and observation apertures are axial and the specimen is moved mechanically in X, Y and optionally Z axes (Sheppard & Wilson, 1978a; Brakenhoff *et al.*, 1986); or (3) the objective lens is moved across a stationary object (Hamilton & Wilson, 1986). Confocal fluorescence microscopes have also been described by Cox (1984), Wijnaendts van Resandt *et al.* (1985), and Takamatsu & Fujita (1988). A confocal module for a scanning fluorescence microscope was described by Shack *et al.* (1987). Confocal fluorescence microscopes are now available commercially from Bio-Rad (Figs. 15.6, 15.7), Leitz, and Zeiss (Oberkochen) (Figs. 15.8, 15.9).

Confocal scanning achieved by moving the specimen, while the illumination and detection systems are focussed at a fixed point on the optical axis, avoids the effects of

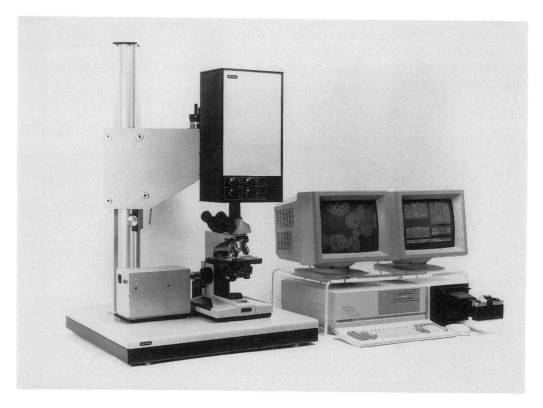

Fig. 15.6. The Bio-Rad/Medical Research Council (MRC) Confocal Conversion System. In this system, the confocal optics, scanning mirrors, dichromatic filters, photodetectors and argon arc laser are housed in a box positioned above the photo tube of a standard microscope (Nikon in this illustration). The signal from the detectors is then processed by the attached personal computer. Of the two video monitors, one displays the normal menus and data from the computer, the other is a high-resolution slave monitor to display the picture. Photograph courtesy of Bio-Rad Laboratories Pty Ltd.

off-axis aberrations in the optical system, and maintains constant image quality from edge to edge of the scanned field, the dimensions of which are limited only by the mechanical stage. The area scanned is independent of the optical magnification. It is also easy to obtain and preserve confocality. On the other hand, scanning by moving the optical images of the source and detector, e.g. by mirrors, allows much faster scanning, and a confocal system can be attached to any microscope without any necessity for a special mechanical stage. Magnification and pixel size are not affected by the limitations of the mechanical stage.

In principle, the excitation beam and the measuring pinhole should be of infinitesimal diameter. In practice, both must be of finite size. The detector pinhole may need to be relatively big in order to achieve sufficient sensitivity to measure weak fluorescence at a reasonable speed. The effects of the finite size of the detection pinhole, and the effect of using a slit, were discussed by Wilson (1989) and Wilson & Carlini (1987, 1989).

The Bio-Rad/Medical Research Council (MRC) Confocal Conversion System (Figs. 15.6, 15.7; White *et al.*, 1987) is a unit which can in principle be attached to any

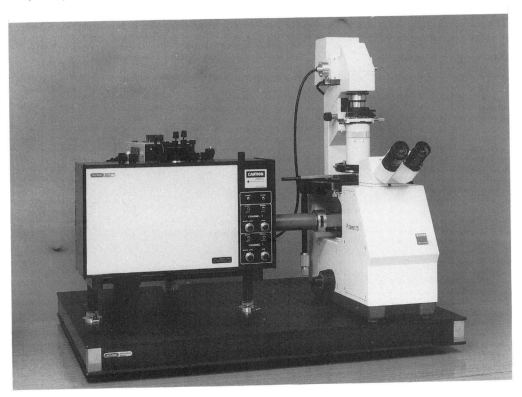

Fig. 15.7. The Bio-Rad MRC Confocal System attached to a Zeiss (Oberkochen) inverted microscope. The scanning system can be attached through either the epi-illuminator or the video port of the microscope. The system illustrated has an additional detector mounted near the lamphousing for scanning transmission microscopy. Photograph courtesy of Bio-Rad Laboratories Pty Ltd.

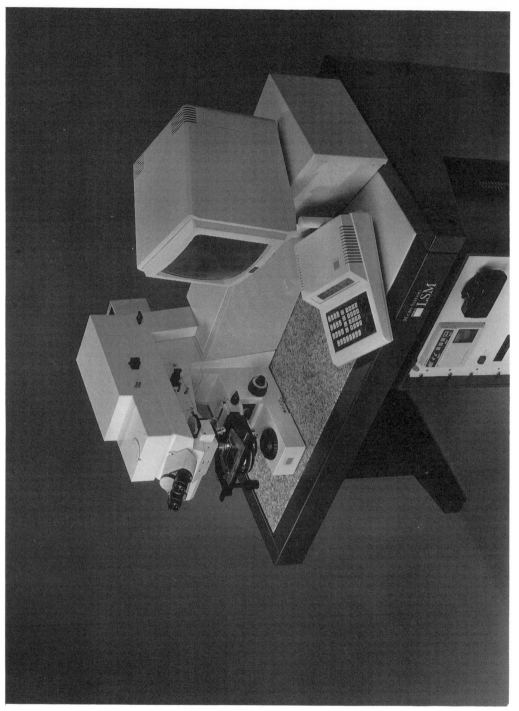

Fig. 15.8. Zeiss LSM laser scanning microscope. Photograph courtesy of Carl Zeiss Pty Ltd, Sydney.

microscope. The device was originally developed at the MRC Laboratory of Molecular Biology, Cambridge, UK. In this system, the confocal optics, scanning mirrors, fluorescence filters and dichromatic mirror, photodetectors and argon ion laser (488 and 514 nm) form a unit which is housed in a box positioned above the photo tube of a standard microscope. Lasers of other wavelengths from UV to IR can be used. The signal from the detectors is processed by a personal computer, which has two video monitors, one which is the normal computer monitor and the other, a high-resolution monitor, to display the picture. One major advantage of this system is that it can be attached to either a conventional vertical microscope (Fig. 15.6) or an inverted microscope (Fig. 15.7); or indeed to any other type of optical microscope.

The Zeiss LSM series (Laser Scan Microscopes; Figs. 15.8, 15.9; Wilke, 1985) are based on a very large stand whose special feature is a wide distance between the objective axis

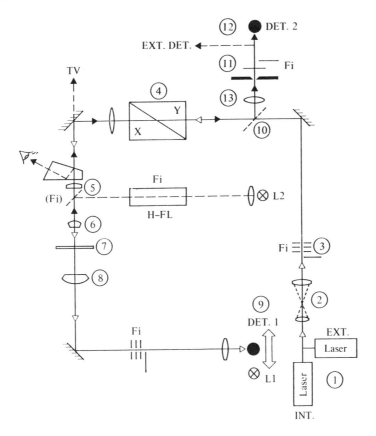

Fig. 15.9. Zeiss LSM laser scanning microscope (schematic). 1, lasers; 2, beam expander; 3, laser light attenuators; 4, X–Y deflection unit; 5, tube lens; 6, objective; 7, specimen; 8, condenser; 9, detector for transmitted light; 10, beamsplitter; 11, barrier filters; 12, detector for fluorescence and reflected light; 13, confocal spatial filter; DET. 1, DET. 2, detectors; EXT. DET, external detector; Fi, filter; H-FL, filter slider with matched excitation and emission filter set; L1, L2, lamps; TV, video camera. Diagram courtesy of Carl Zeiss Pty Ltd, Sydney.

and the stand, so that large objects can be accommodated. Several models already exist, for various applications in biology, medicine, and materials sciences. Both confocal and line scan operating modes are provided. For fluorescence, standard excitation wavelengths are 488 nm and 514 nm (argon laser) and 633 nm (helium-neon laser), supplemented by a conventional fluorescence epi-illumination system with 50 W mercury arc lamp. Software enables images to be created showing sections, 3-D simulations, or stereo pairs, and processed in a variety of ways.

Leitz

The confocal laser scanning microscope of Wild Leitz is based on the Leitz Diaplan stand, equipped with a special intermediate tube for coupling an argon laser to the scanning and detection unit. Images are stored in a 256×256 pixel format. Sections and stereo pairs can be produced by software, which has a large number of image-processing routines.

Applications

Applications of confocal fluorescence microscopy include studies of dental enamel (Boyde, 1987), the location of microspheres in phagocytosis research (Hook & Odeyale, 1989), organization of DNA (Arndt-Jovin, Robert-Nicaud & Jovin, 1990), RNA hybridization (Bauman, Bayer & van Dekken, 1990), and microstructure of solid industrial materials (Knoester & Brakenhoff, 1990). The unitary-beam type of CSFM lends itself to studies of fluorescence recovery after photobleaching (Scholz, Grosse-Johannböcke & Peters, 1988). Other applications, using the tandem scanning system, were listed by Boyde & Watson (1989).

IMAGE PROCESSING

Digital processing

The video signal for a picture can be digitized, i.e. converted to an array of numbers each representing the light intensity at a particular spot (i.e. the brightness of a particular pixel). The image is usually digitized in a matrix of 512×512 pixels. The picture can then be stored in computer memory, recorded as computer data (e.g. on magnetic tape or a floppy disk) and displayed on a computer monitor or printed with a digital printer (laser printer), and subjected to analysis by computer. The techniques for acquisition, processing and analysis of digital images are now well advanced due to developments as part of the United States' space research program. For reviews of this technique applied to microscopy see Arndt-Jovin *et al.* (1985) and Jarvis (1988). For more general reviews, see Watkinson (1988), Wahl (1987), Green (1983), Talmi (1983), Ballard & Brown (1982), Pavlidis (1982), Rosenfeld & Kak (1982), and Serra (1982).

Image analysis of the fluorescence image enables the correlation of morphological and quantitative data. Quantitative fluorescence image analysis was described by Takamatsu, Kitamura & Fujita (1986). Image analysis of microscope images in general was reviewed by Jarvis (1988) and Serra (1986); see also Serra (1982) and van der Ploeg *et al.* (1977a), and the general reviews cited in the previous paragraph.

Digital enhancement techniques involve computer processing of a digitized image, in order to optimize the image in respect of specified criteria. Given a rapid digitizer and a sufficiently fast computer, manipulation is possible in real time to permit continuous observation. Possible changes to the image include: (1) expansion or contraction of the contrast range, by manipulation of the grey scale (Inoué, 1981; Sahota *et al.*, 1981); (2) the sharpening of edges (Sahota *et al.*, 1981; Walter & Berns, 1981); (3) reduction of random video noise by averaging several images (Steponkus *et al.*, 1984); and (4) displaying the difference between images to remove fixed patterns and show changes (Allen & Allen, 1983). Digital enhancement procedures can be applied to any digitized image, e.g. one obtained with excitation scanning.

Analogue processing

Analogue devices can be used for image processing, and may be faster in operation and cheaper to assemble. Burton & Bank (1986) described a videomicroscope system with control of gain and contrast ranges of the video signal, using commercially available analogue circuits; edge enhancement was obtained by selective amplification of abrupt voltage changes in the signal, and fine detail was enhanced by boosting high frequency components of the video signal.

The Joyce-Loebl MagiCal

The Joyce-Loebl MagiCal is an image analysis system for studying quantitatively the intracellular distribution of ions such as Ca^{2+}, Na^+ and H^+, and temporal changes in their concentration and distribution in response to stimuli. The name 'MagiCal' refers specifically to image analysis of calcium. The system consists basically of a fluorescence microscope (such as the Nikon Diaphot, see Fig. 2.5), a highly sensitive video camera, a computerized image analysis unit and a computerized control system.

Because probes for ionic concentration require measurement at two wavelengths (see Chapter 10), the system incorporates motorized filter changers in both excitation and emission pathways. These consist of rotating discs, each holding four filters; the disc can either be stepped from one filter to the next when required, or rotate continuously in synchrony with the 25 frame/s video rate so that each successive video frame is made through a different filter. Four filters allow, in principle, for use of two pairs of filters for concurrent observation of different ions, e.g. Ca^{2+}, and pH. Otherwise, the two pairs are normally identical, for the same fluorophore.

Sequences of images are captured at normal video rate, at wavelengths appropriate for the fluorochrome being used. The images are captured in user-selectable sizes from 64×64 pixels to 512×512 pixels, digitized into 64 or 256 grey levels. Capture is controlled by automatic internal or external timing, or manually by pressing keys. The interval between successive frames can be varied from 80 ms to several hours. Up to 32 megabytes of digital storage is provided within the image analysis unit; additional storage can be added by using an external hard disc, or the data can be dumped to any standard digital storage device (such as tape, compact disc, etc). The image analyser unit uses Joyce-Loebl software, which can be modified by the user if desired, programming in PASCAL.

The image analysis unit rapidly calculates the ratios of corresponding pixels in pairs of images taken at different wavelengths. Further image processing can be carried out, including averaging or integrating successive images, correction for background, and measurements of individual structures.

16

The history of quantitative fluorescence microscopy

This chapter is intended to give a general outline of the historical development of the instrumentation, techniques and major applications of quantitative fluorescence microscopy. Reviews of the history of fluorescence microscopy include those of Haitinger (1938, 1959), Ellinger (1940), de Lerma (1958), Kasten (1983, 1989), and Rost (1990). Recent developments have been reviewed by Tanke (1989). There does not appear to have been any previous major review of the history of quantitative fluorescence microscopy. Papers now of significant historical importance include those of Mellors & Silver (1951), Nordén (1953), and Rigler (1966).

There are four main threads to be followed in the development of the techniques described in this volume: the development of analytical techniques by microspectrofluorometry, the development of quantification by microfluorometry, the extension of measurements to other parameters such as polarization and time resolution, and the application of techniques for scanning.

Because fluorescence microscopy was first developed mainly for studying intracellular phenomena by means of tracers, the establishment of fluorescence microscopy at a qualitative level was fairly rapidly followed by the development of analytical microspectrofluorometry to enable the identification of fluorophores in cells. Microspectrofluorometry was first developed by Policard & Paillot (1925) and Borst & Königsdorfer (1929), who added a monochromator on the emission side of a fluorescence microscope. However, the main groundwork of microspectrofluorometry was laid by Nordén (1953), and the firm establishment of microspectrofluorometry as a routine tool is due largely to the work of Caspersson's team at the Karolinska Institute (Caspersson, Lomakka & Rigler, 1965; Jonsson & Ritzén, 1966; Rigler, 1966; Ritzén, 1967; Caspersson & Lomakka, 1970).

The introduction of scanning was stimulated by the application of Acridine Orange staining to the cytological diagnosis of cancer. Mellors & Silver (1951) were the pioneers in this, using a mechanical scanner. The further development of scanning techniques had to wait for the prerequisite development of electronic technology, particularly television and computerized image analysis, as described in Chapter 15.

Quantification by microfluorometry presents numerous difficulties because of the various sources of error which I discussed in Chapter 4. The development of this technique was therefore comparatively slow. It was Fritz Ruch who was the main pioneer in this field, starting with the construction of a microfluorometer (Ruch & Bosshard, 1963), and developing techniques (Ruch, 1964, 1970; Ruch & Leemann, 1973) and applications, particularly to the measurement of DNA (Ruch, 1966c, 1973). In commercial development, a notable contribution was the Leitz MPV photometer, whose design was particularly suitable for application to microfluorometry.

Fluorescence microscopy was discovered as part of a search by Köhler for a form of microscopy which increased resolution. It was proposed to increase the resolution in accordance with the Abbé theory by reducing the wavelength. Since that original discovery the wavelength used for excitation has been getting longer, progressing from the ultraviolet (Köhler), to the violet and short-wavelength blue (Ziegenspeck, 1949), the green for rhodamines and now most recently the red for green dyes. The use of green dyes, implying as it does emission in the infrared, requires electronic or photographic assistance for visualizing the fluorescence. This continuing progression towards longer wavelengths for both excitation and emission, while offering advantages in reduced damage to the specimen, has brought about a corresponding degradation in the possible resolution. The situation has been greatly improved by the use of scanning at the excitation wavelength, so that the resolution is based on the excitation rather than the emission wavelength, and by the use of confocal microscopy, which increases the resolution beyond the Abbé limit. Recent developments in scanning microscopy seem likely to lead to even greater increases in resolution.

Early microfluorometers

Numerous microfluorometers have been described in the literature. The first generation of these employed, as sources of monochromatic light for excitation, arc lamps and filters; the second generation employ lasers, and are described in Chapter 2.

Arc-and-filter microfluorometers which have been described include those of Mellors & Silver (1951), Mellors, Glassmar & Papanicolaou (1952), Chance & Legallais (1959), Goldman (1960), Eränkö & Räisänen (1961), Donáth (1963), Ruch & Bosshard (1963), Bosshard (1964), Hovnanian, Brennan & Botan (1964), Kunz (1964), Mansberg & Kusnetz (1966), Thaer (1966a), van Gijzel (1966), Goldman (1967), Böhm & Sprenger (1968), Stoward (1968a), Sprenger & Böhm (1971a), Taylor, Heimer & Lidwell (1971), Ruch & Trapp (1972), Combs (1973), Enerbäck & Johansson (1973), Wasmund & Nickel (1973), Haaijman & Wijnants (1975), Teichmüller & Wolf (1977), the Cutler-Hammer Corporation (Barden, Aviles & Rivers, 1979), de Josselin de Jong, Jongkind & Ywema (1980), Ji et al. (1979), Cowell & Franks (1980), Phillips & Martin (1982), and Nobiling & Bührle (1989). Instrumentation up to about 1965 was reviewed by Thaer (1966a).

Most of these instruments were designed for specific applications involving a single fluorophore, requiring only a single excitation wavelength. Apart from questions of stabilization, this limited requirement for excitation permitted the use of a high-pressure mercury arc lamp, the 365 nm line being usually selected by means of a glass filter (which latter is perfectly satisfactory for the purpose). In most cases, the secondary filter was of the barrier type, passing all radiation above a given wavelength, rather than a narrow-band filter. This arrangement gives increased sensitivity and simplicity at the expense of optical specificity.

Epi-illumination

In early instruments, darkground condensers were used for illumination, in order to minimize background intensity, although there is difficulty in obtaining even illumination of the field; a few workers preferred brightfield substage condensers. Epi-illumination (vertically incident exciting light) was developed quite early; the first users appear to have been Policard & Paillot (1925) in France, who examined the fluorescence on the surface of tissues using a Greenough (stereo) microscope. However, Ellinger & Hirt (1929a,b, 1930) were the first to develop a system for general fluorescence microscopy. The stimulus for this development was the need for microscopy of living organisms. Even with a conventional microscope, owing to the size and opacity of the object, it had already become necessary to use epi-illumination. Depending on the object, much of the incident light was either reflected from the surface or absorbed by superficial layers of the object, being thereby lost for the formation of the microscope image. The newly developed fluorescence microscopy with fluorochromes, used in conjunction with epi-illumination, was ideal since incident light reflected from the surface of the specimen is absorbed by the barrier filter and therefore does not degrade the image; moreover fluorescence can display small intracellular particles. A fluorescence microscope with epi-illumination, intended for intravital microscopy, was produced by Carl Zeiss to the specifications of Ellinger & Hirt (1929b, 1930). This was followed by a number of similar instruments by other firms: Leitz; Bausch & Lomb instructed by Singer (1932); and C. Reichert instructed by Pick (Mehler & Pick, 1932; Pick, 1934), unfortunately neglecting a number of important features in the original method (Ellinger, 1940). A Leitz fluorescence epi-illumination system was illustrated by Haitinger (1938).

A major advance in epi-illumination was the introduction of a dichromatic mirror, reflecting more than transmitting at the excitation wavelength and conversely transmitting more than reflecting at the emission wavelength. This kind of reflector has now become standard in epi-illumination systems. It was first considered by Mellors & Silver (1951), who, in the process of developing a microfluorometer for scanning cancer cells (see Chapter 15), found difficulty in obtaining suitable filter combinations for fluorochromes emitting in the green and blue regions. For that and other reasons they investigated the potentialities of vertical reflected illumination using a dichromatic mirror with high reflectance in the ultraviolet (UV) and low reflectance in the visible, with reciprocal transmission characteristics. Apparently the mirror proved to be satisfactory, but nothing further appears to have been published about their experiments. It was Ploem (1967) who rediscovered the principle, and was responsible for the commercial application of dichromatic mirrors in epi-illuminators.

Rigler (1966) recognized the advantage of epi-illumination in reducing absorption error in microfluorometry. Rigler, using the Zeiss–Caspersson UMSP described below, did not have the appropriate equipment, and the first practical development of epi-illumination for microfluorometry and microspectrofluorometry was carried out by me (Pearse & Rost, 1969; Rost & Pearse, 1971).

Other features

Inverted microscopes for microfluorometry were introduced by Chance & Legallais (1959); their instrument was subsequently modified for epi-illumination to facilitate micromanipulation (Kohen & Legallais, 1965; Kohen, Kohen & Thorell, 1969), and a sophisticated instrument, with a multichannel detector, was described in detail by Hirschberg *et al.* (1979). These instruments are used for studies on living tissues.

A reference channel (to compensate for variations in intensity of excitation) was incorporated into the instruments of Mellors *et al.* (1952), Chance & Legallais (1959), Ruch & Bosshard (1963), van Gijzel (1966), and Goldman (1967).

The detector has usually been a photomultiplier. Early workers used RCA 1P21 tubes having an S-4 cathode; there has been subsequent progress to S-11 and S-20 cathodes, which have a greater sensitivity to red. Other means of detection have been photography (Eränkö & Räisänen, 1961), a photocell (Donáth, 1963), the Berek visual comparison photometer (Ruch & Bosshard, 1963; van Gijzel, 1966), and a cadmium sulphide photoresistor (Stoward, 1968a).

Commercial microfluorometers

Equipment for microfluorometry has been available commercially for quite some years. Leitz, in particular, appear to have paid attention to the development of equipment for microfluorometry. Independent companies also supply some equipment which can be used in conjunction with a microscope.

Leitz have made several photometer units over the years, in the modular MPV system (the acronym MPV stands for *Mikrophotometer mit Variable Messblende*, signifying an advance over the earliest model which had fixed field diaphragms). The MPV-1 is illustrated in Figs. 13.1 and 16.3. It consisted of a module which attached at the top of a trinocular tube (in place of a camera) and was normally supported by a bracket from the rear of the microscope stand. It contained, essentially, a measuring field diaphragm, a viewing arrangement, a filter for monochromation, and a photomultiplier tube. The measuring field diaphragm was interchangeable; those normally supplied were circular (iris) and rectangular. I found it useful to have a holder made, to accept interchangeable circular apertures of fixed diameter. The instrument included an arrangement whereby an image of the object, together with a superimposed image of the field diaphragm, could be seen in a side tube. The side tube normally was fitted with a Leica camera and reflex viewing system (Fig. 16.2; the same photographic system is also shown in Fig. 16.4), but instead I usually use a simple tube with an eyepiece (Fig. 13.1). A range of interchangeable filters were supplied by Leitz; I had a holder made to accept small (25 mm square) interference filters. The photomultiplier tube was situated at the top of the apparatus, and was connected to an appropriate power supply and meter for the photomultiplier. The application of the MPV-1 to microfluorometry was described by Thaer (1966a).

Reichert (C. Reichert Optische Werke A.G.) made a microscope photometer (Gabler *et al.*, 1960) which could be used on their Zetopan fluorescence microscope. Light for excitation was obtained from a mercury arc lamp, using filters. Emission was measured from an area determined by fixed circular apertures, monochromated by an interference wedge filter. The equipment has been used by Kunz (1964). A more sensitive version was described by Nairn *et al.* (1969) for quantification of immunofluorescence.

Zeiss (Oberkochen) have made several microphotometers, and Zeiss microscopes have been the basis of a number of independently built devices. The Zeiss MPM, equipped for microfluorometry, was released in 1967. This was superseded in 1970 by the MPM 01 (Fig. 16.1) which has only recently been discontinued. The Zeiss Microscope Photometer SF was a simple device adaptable to the Universal, Universal R, ACM with tube head, and the Photomicroscope. It consisted of little more than a photo-multiplier housing with a measuring field diaphragm, and was intended as a relatively inexpensive instrument for answering simple questions.

The Zeiss Cytofluorometer (Ruch & Trapp, 1972) was based on a standard microscope (Zeiss Universal or Photomicroscope) with the photometer MPM 01. The

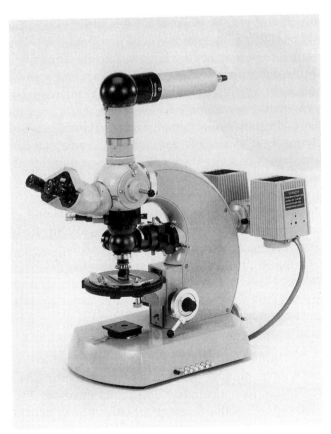

Fig. 16.1. Zeiss MPM 01 photometer system, for microfluorometry on the Axioplan microscope. Photograph courtesy of Carl Zeiss Pty Ltd, Sydney.

photometer unit contained interchangeable field diaphragms, a quartz field lens for pupil imaging, a moveable mirror, and a photomultiplier. Depending on the position of the mirror, the object image was directed forwards into an eyepiece, upwards to the photomultiplier, or backwards to a photographic or video camera. An electrically controlled shutter could limit fluorescence excitation during measurement to about 7 ms to reduce fading. The measuring sequence was controlled electrically from a foot switch.

Cowell & Franks (1980) described a modified Zeiss microfluorometer based on a Zeiss Photomicroscope with a Zeiss MPM 01 photometer head and a stabilized HBO 100 mercury arc lamp. The measuring system consisted of a photometer with RCA type 931 photomultiplier, an amplifier, a peak detector, a printer supplied by Bentham Instrument Limited (Reading, UK) and a shutter switching system. The illuminated field was limited by a pinhole diaphragm to avoid irradiation of surrounding objects. They described the use of the instrument for quantification of DNA in nuclei.

Microspectrofluorometers

The earliest construction and use of a microspectrofluorometer appear to have been those of Policard & Paillot (1925), Borst & Königsdorfer (1929) and Nordén (1953). The latter's discussion of the principles involved is still relevant. His paper laid down most of the theoretical requirements for microspectrofluorometry, omitting only the advantages of epi-illumination (discussed by Rigler, 1966) and photon counting (applied by Pearse & Rost, 1969). The first really effective microspectrofluorometer was devised by Caspersson and his colleagues (1965; Fig. 16.1), as a modification of the Zeiss (Oberkochen) Ultramicrospectrophotometer (UMSP) which had also been devised in that laboratory (Caspersson et al., 1953). Commercial developments were a little slow to follow; however, Leitz took a special interest in this area.

Instruments have been described by many authors, notably Policard & Paillot (1925), Borst & Königsdorfer (1929), Nordén (1953), Rousseau (1957), Olson (1960), West (1965), Caspersson et al. (1965), Runge (1966), Thieme (1966), Björklund, Ehinger & Falck (1968b), Parker (1969b), Van Orden (1970), Rost & Pearse (1971), Combs (1973), Cova, Prenna & Mazzini (1974), Mayer & Novacek (1974), Ploem et al. (1974), David & Galbraith, (1975) Wreford & Schofield (1975), Boldt et al. (1980), Quaglia et al. (1982) and Ghetti et al. (1985). Except for the early instruments up to that described by Rousseau (1957), these instruments all had proper monochromators for both excitation and emission; a number of other instruments have been described with a filter for excitation and a monochromator for emission analysis (Thaer, 1966b; Jotz, Gill & Davis, 1976; Ottenjann, 1980).

Nordén (1953) used a prism monochromator for emission measurement and filters for excitation. Measurements were made in a fixed area of about 23 μm diameter. The instrument was used for spectrofluorometric identification of benz[a]pyrene in tissues.

Rousseau (1957) briefly described an instrument capable of photographically recording emission spectra using the mercury 365 nm line for excitation. Accuracy was stated not to exceed 20%.

Olson (1960) developed a microspectrofluorometer based on a Reichert MeF inverted microscope, with capability for rapid scanning of emission spectra for the study of the photochemistry of living cells. The fluorescence emission was dispersed by a Zeiss spectrometer ocular and scanned by a revolving spiral slit. The light was measured by a photomultiplier, and the output displayed on an oscilloscope. At the end of each wavelength scan, light from a reference channel generated by a glass beamsplitter was sampled by a prism mounted on the scanning disc, so that the stability of the system could be assessed by the height of the blip produced on the oscilloscope screen. By a modification of the scanning system, it was possible to record time changes in fluorescence intensity at several wavelengths and also absorption at the excitation wavelength.

West, Loeser & Schoenberg (1960) applied television techniques to the recording of emission spectra. The emission spectrum was dispersed by a Leitz prism monochromator, the exit slit of which was removed. The spectrum so produced was scanned by an RCA Intensifier Image Orthicon television camera, and displayed as a graph on an oscilloscope screen by the technique of line selection (Loeser & Berkley, 1954). This instrument had the advantages of great sensitivity and rapid scanning of emission spectra. This instrument was later modified (West, 1965) by the additions of a prism monochromator for excitation, and of a heating stage.

The Zeiss–Caspersson Ultramicrospectrophotometer (UMSP; Caspersson *et al.*, 1953; Caspersson & Lomakka, 1970) was modified as an important instrument for fluorescence studies (Caspersson *et al.*, 1965; Caspersson & Lomakka, 1970). A view of it is shown in Fig. 16.2. This instrument was further described by Rigler (1966) and Ritzén (1967). The principle of the machine is as follows. Excitation was by a 500 W xenon arc lamp and a standard Zeiss quartz-prism monochromator using dia-illumination with an achromatic UV condenser or a darkground condenser. An identical second monochromator selected the wavelength at which emission was measured. For scanning spectra, the wavelength drum of either monochromator could be rotated continuously by a motor. The measuring field was determined by a choice of iris, hole and slit diaphragms, corresponding to measurement areas down to 0.3μm in diameter. The light beam was chopped by a rotating sector shutter, allowing the photomultiplier to measure alternately the light from the second monochromator or from a reference light path. In the reference beam, a cuvette containing a solution of Rhodamine B was used as a quantum converter, converting UV to visible light. The photomultiplier signal was amplified, and the ratio of the outputs from the two channels (microscope and reference) was recorded on a chart recorder. A built-in polarizer and analyser allowed the determination of the degree of polarization of emitted light.

Runge (1966) described a microspectrofluorometer designed particularly for the study of porphyrin fluorescence spectra. Excitation was by means of a current- or voltage-stabilized mercury arc lamp (HBO 200). The excitation wavelength was selected by filters. Emission wavelength selection was by a Steinheil microspectrograph, with photomultiplier detection. Output from the photomultiplier was taken by a tuned inductance-capacitance anode load, resonating with a mechanical light chopper.

Thieme (1966) described a versatile microspectrofluorometer, in which both excitation and emission wavelengths were selected by grating monochromators. The monochromator drums were rotatable by synchronous motors, operable either singly or together: the latter arrangement were used for absorption measurements. Each monochromator was connected to a variable voltage divider giving an analogue signal related to the wavelength, to drive one axis of an $X-Y$ chart recorder. Illumination of the specimen was from below by a darkground condenser. The measuring field was determined by a diaphragm acting also as an entrance aperture to the second monochromator. Measurement could be made in an area 1 μm diameter. The detector was a photomultiplier type 1P21, connected to an $X-Y$ recorder with a greatest sensitivity of 0·1 nA full-scale. No provision was made for monitoring the light source.

Van Orden (1970) at the University of Iowa described a microspectrofluorometer based on the Leitz Orthoplan microscope with MPV-1 photometer attachment (Fig. 16.3). The illumination system included a Hanovia 1000 W xenon arc lamp, a Schoeffel quartz prism monochromator, and a substage darkground condenser. The measuring system consisted of an MPV-1 photometer, with a second Schoeffel quartz prism monochromator interposed between the body of the MPV and the photomultiplier, an EMI D-224B (a small-cathode version of the 9558). No reference channel was

Fig. 16.2. The Zeiss–Caspersson Ultramicrospectrophotometer (UMSP) as modified for fluorescence (Caspersson, Lomakka & Rigler, 1965), shown operated by Dr Martin Ritzén. The two chart recorders are for recording excitation and emission spectra. Institute for Cell Research and Genetics, Karolinska Institute, Stockholm. Photograph taken by me in 1966. Thanks are due to Professor T. Caspersson and Dr M. Ritzén for permission to take this picture.

employed. This instrument became the prototype of several similar instruments, notably that at Monash University, which was further developed to provide automatic computerized correction of spectra (Wreford & Schofield, 1975; Wreford & Smith, 1982).

Parker (1969b) described a microspectrofluorometer formed by combining a conventional microscope (Gillett & Sibert) with a spectrophosphorometer. The resulting instrument was unique in that it could be used for both microspectrofluorometry and, by means of mechanical choppers in the light beams, for microspectrophosphorometry (see Chapter 12). Excitation was obtained by mercury or xenon arc lamps through a Bausch & Lomb grating monochromator. The emission spectrum was measured through a Hilger D-284 quartz-prism double monochromator with an EMI 9558

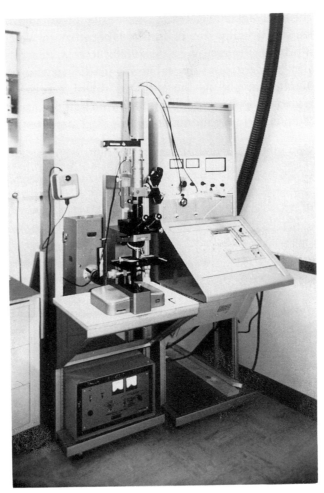

Fig. 16.3. The Leitz-Schoeffel microspectrofluorometer of Van Orden (1970), Oakdale Toxicology Center, University of Iowa, USA., photographed in 1968, by permission of Dr L.S. Van Orden. The instrument is based on a Leitz Orthoplan microscope, with MPV-1 photometer, and Schoeffel monochromators.

photomultiplier (S-20 cathode). A reference channel could be employed, with a fluorescent-screen quantum converter.

Commercial microspectrofluorometers
Earlier instruments for microspectrofluorometry, with monochromators for both excitation and emission, which were commercially available, included (as far as I know, in chronological order of appearance): the Leitz microspectrograph modified for fluorescence, the Leitz–Schoeffel microspectrofluorometer, and the Farrand Microscope Spectrum Analyser (MSA).

The Leitz microspectrograph was originally designed by Ruch (1960) for measurements of UV absorption. The instrument was later available in modified forms for microspectrofluorometry (Figs. 16.4–16.7). Of the commercial microspectrofluorometers, this was in its day the most advanced. However, each of the several instruments appears to have been modified individually for, or in, the laboratory in which it was installed (Thaer, 1966b; Björklund, Ehinger & Falck, 1968b; Rost & Pearse, 1971).

In the original instrument, an adjustable slit in the viewing system acted as a measuring field diaphragm and as the entrance slit of a prism spectrograph; the spectrum was recorded on a photographic plate, which might later be scanned with a densitometer. A standard Leitz monochromator was used to monochromate the

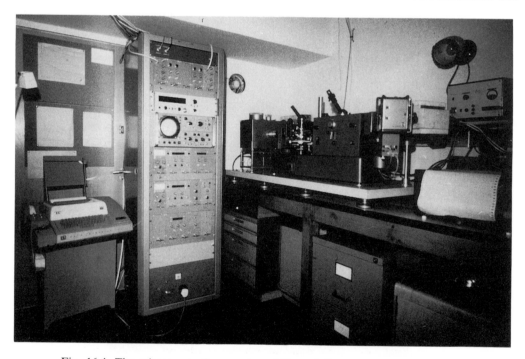

Fig. 16.4. The microspectrofluorometer of Rost & Pearse (1971). This was a modified Leitz microspectrograph. It had epi-illumination, photon counting, and a digital data logging system. Left to right: Teletype machine with keyboard for entering data; power supply and photon-counting equipment; and microspectrofluorometer (detail shown in Fig. 16.5).

excitation. Because of the length of the photographic exposure, considerable fading of the fluorescence occurred. To obviate this difficulty, Thaer (1966b) used a television camera to scan the spectrum. Using line selection technique, a graph of the spectrum could be displayed on a storage oscilloscope. With this instrument, Thaer (1966b) was able to show changes in the fluorescence emission spectrum of Acridine Orange during fading. Subsequently, the television camera was replaced by a photomultiplier behind a slit, with an oscillating mirror by means of which the emission spectrum was scanned.

Our own instrument (Fig. 16.4; Rost & Pearse, 1968, 1971; Pearse & Rost, 1969; Rost, 1973) at the Royal Postgraduate Medical School, London, was the first of these to take full advantage of the possibilities of epi-illumination at all wavelengths, a reference channel for correction of excitation spectra, and photon counting. It was also unique in possessing a third channel for simultaneous determination of absorption at the excitation wavelength. Excitation was obtained with either a mercury arc or xenon arc lamp (for emission and excitation spectra respectively), and a standard Leitz prism monochromator. An epi-illuminator produced to my specifications, employing a quartz plate as beamsplitter, and Zeiss Ultrafluar objectives permitted excitation with

Fig. 16.5. Detail of the microspectrofluorometer of Rost & Pearse (1971). Left: the illumination unit, with (left to right), the lamphouse; a mirrorhouse with shutter; a monochromator, raised up on a stand to permit epi-illumination (stand is removed for dia-illumination); and quartz projection lens (power supply for lamp behind). Right: the microspectrograph unit, showing upright carrying leaf shutter and field diaphragm, microscope with epi-illuminator, part of the emission dispersing and measuring unit, and camera.

epi-illumination well into the UV region. A reference channel was constructed using another quartz beamsplitter (made from a standard quartz coverslip), a quartz cuvette containing Rhodamine B solution, and a photomultiplier system. Photon-counting systems were used for both photomultipliers (main and reference channels), with automatic digital data logging.

Other modifications of the Leitz microspectrofluorometer have been described by Björklund, Ehinger & Falck (1968b), and by Sprenger & Böhm (1971a). The instrument of Björklund, Ehinger & Falck (1968b) uses a quantum converter similar to that described by Ritzén (1967) for checking the intensity of the excitation beam. The instrument of Sprenger & Böhm (1971a) does not have a monochromator for excitation, being intended mainly for quantification of Feulgen reactions. Similar instruments, special modifications of the microspectrograph, were developed in the laboratories of Ruch (Zurich; Fig. 16.6), Prenna (Pavia; Fig. 6.7) and Ploem (Leiden).

Schoeffel (Schoeffel Instrument Corporation, Westwood, NJ, USA; Schoeffel Instrument GmbH, Trappenkamp, Germany) marketed a commercial version of the Van Orden microspectrofluorometer (Van Orden *et al.*, 1967; Van Orden, 1970). This consisted of a Leitz Orthoplan fluorescence microscope, with standard MPV photometer, together with a Schoeffel xenon arc lamp system (1000 W) and two Schoeffel quartz-prism monochromators, one for excitation and the other incorporated into the MPV system. In the prototype, only dia-illumination was used, but epi-illumination is equally practicable. The instrument is much improved by the addition of stray-light

Fig. 16.6. The Ruch microspectrograph modified for fluorescence. Photograph taken in 1969, in the Department of General Botany, Eigenossische Technische Hochschule, Zurich, with permission of Professor F. Ruch.

filters after the excitation monochromator. The main users of this instrument were Van Orden (Van Orden *et al.*, 1967; Van Orden, 1970) and Wreford (Wreford & Schofield, 1975; Wreford & Smith, 1982).

Farrand (Farrand Optical Co., Inc., New York, USA) produced an instrument called the Microscope Spectrum Analyser (MSA), which consisted, essentially, of the components of a microspectrofluorometer without the microscope, the latter being provided by the individual customer. This system could be adapted, in principle, to any microscope suitable for fluorescence microscopy and having a monocular tube to accept the eyepiece of the MSA. It was effective, and the least expensive of the microspectrofluorometric systems available at that time.

The MSA consisted of two independent units, respectively for excitation and for measurement of emission. The excitation unit consisted of a stabilized arc lamp, with a grating monochromator. The measuring unit consisted of an eyepiece, with provision for diverting the light from the centre of the field through a hole acting both as field diaphragm and as entrance aperture of a small monochromator. This aperture was fixed but interchangeable with apertures of other sizes; the corresponding area was indicated in the eyepiece by an image of the aperture. On the output side of the monochromator was a conventional photometer system using a photomultiplier. The MSA system had an obvious appeal because of its simplicity and economy. Mounted on a suitable microscope stand equipped with an epi-illumination system and a monocular or 'photographic' tube to accommodate the measuring eyepiece, the MSA

Fig. 16.7. The microspectrofluorometer, a modified Leitz microspectrograph, in the Centro di Studio per l'Istochimica del CNR, University of Pavia. Photograph taken in 1987, with permission of Professor G. Bottiroli.

seemed to be particularly suitable for answering simple questions in the 360–700 nm spectral region, e.g. for determining the spectra of fluorochromes, or for differentiation between catecholamines and serotonin by formaldehyde-induced fluorescence.

Gamma made a photometric attachment for microscopes, which has been used in a microfluorometer by Taylor *et al.* (1971). The Gamma attachment consisted of an eyepiece with attached photometer; light was led from a small area at the centre of the field by a fibre-optic to a photometer.

Applications

Microfluorometry was applied particularly to the measurement of nuclear DNA using a fluorescent Schiff-type reagent in the Feulgen method. Prominent in this field were Giovanni Prenna (in Pavia, Italy), and Fritz Ruch (in Zurich, Switzerland), who developed both instrumentation and techniques.

The electronic era

As in many other fields, quantitative fluorescence microscopy has been revolutionized by the past half century's developments in electronics and opto-electronic devices. Although quantitative fluorescence microscopy has been carried out using the Berek visual comparator for the measurement of brightness (Ruch & Bosshard, 1963; van Gijzel, 1966) and photography has been used for quantification (Eränkö & Räisänen, 1961) and recording of emission spectra (Policard & Paillot, 1925; Borst & Königs-dorfer, 1929; Rousseau, 1957), yet the development of microfluorometry was made possible by electrical measuring devices. Of these, the photomultiplier was at first by far the most important. This requires a high-voltage power supply but its output can if necessary be measured with quite simple devices. The application of photon counting, using circuitry originally developed for scintillation counting, and keeping the photo-multiplier cool and dry, gave substantial increases in the sensitivity of the photomultip-liers. High-speed photomultipliers and associated circuitry enable studies of time resolution.

Television was first used for scanning emission spectra in microspectrofluorometry (West *et al.*, 1960; Thaer, 1966b). The most recent generation of video cameras using charge-coupled device (CCD) sensors are sufficiently sensitive and sufficiently stable to enable fluorescent images to be scanned and measured.

Computers from the start facilitated the processing of data. Electronic devices could be built to control the operation of shutters and data acquisition. However, it is now possible, in instruments such as the 'MagiCal', for a computer to control the operation of an instrument, the acquisition of data from the video camera, and the storage and subsequent analysis and presentation of such data. The possibility of assembling a scanned image as a video frame and its presentation on a video monitor via a computer greatly increases the versatility of present-day apparatus. It is the method most commonly used for confocal fluorescence microscopy.

It is obvious from the above that major developments in quantitative fluorescence microscopy during the past 50 years or thereabouts have for the most part been concerned with the application of opto-electronics for increasing the sensitivity, speed and versatility of the instrumentation.

Recent developments

The possible extension of microfluorometry to microphosphorometry was realized by Parker (1969b), who was studying the uptake by marine plankton of oil spillage. More recently, quantitative studies have investigated other fluorescence parameters such as polarization and decay time. Flow cytometry enables rapid studies on cell populations, giving results which can be correlated with fluorescence microscopic observations.

Lasers were first applied to microfluorometry by Kaufman, Nester & Wasserman (1971); this was shortly followed by the use of lasers for scanning (Slomba *et al.*, 1972). Tuneable lasers for the measurement of excitation spectra were introduced by Quaglia *et al.* (1982).

Scanning

Scanning of fluorescence has been used for quite some time. Mellors & Silver (1951), Mansberg & Kusnetz (1966), and Gillis *et al.* (1966) used event-counting systems for scanning (see Chapter 15).

Video fluorescence microscopy has brought about a resurgence of activity in studies of intracellular distribution of fluorescent probes in living cells, because of the extreme sensitivity of current devices and the possibilities of making recordings for study at leisure. Quantitative video fluorescence microscopy enables the correlation of phenomena in various parts of cells.

Confocal fluorescence microscopy. It is fitting that this chapter should end with a reference to confocal scanning fluorescence microscopy (described in Chapter 15), which has been hailed as 'arguably the most significant advance in biological light microscopy in this decade' (Editorial comment, *Proceedings of the Royal Microscopical Society*, 1988, p. 289). This technique produces precise optical sections, free from blur and reduction of contrast by fluorescence from out-of-focus planes, and with potentially greatly increased resolution. The system also reduces fading (by limiting the area irradiated at any moment) and is applicable to emission in the UV and infrared, the latter with no loss of resolution because the resolution is that of the excitation and not of the emission wavelength.

Appendix 1

Conversion table, photon energy (eV) to wavelength (nm)

	·00	·01	·02	·03	·04	·05	·06	·07	·08	·09
1·5	827	821	816	810	805	800	795	790	785	780
1·6	775	770	765	761	756	751	747	742	738	734
1·7	729	725	721	717	713	708	704	700	697	693
1·8	689	685	681	677	674	670	667	663	659	656
1·9	653	649	646	642	639	636	633	629	626	623
2·0	620	617	614	611	608	605	602	599	596	593
2·1	590	588	585	582	579	577	574	571	569	566
2·2	564	561	558	556	553	551	549	546	544	541
2·3	539	537	534	532	530	528	525	523	521	519
2·4	517	514	512	510	508	506	504	502	500	498
2·5	496	494	492	490	488	486	484	482	481	479
2·6	477	475	473	471	470	468	466	464	463	461
2·7	459	457	456	454	452	451	449	448	446	444
2·8	443	441	440	438	437	435	433	432	430	429
2·9	428	426	425	423	422	420	419	417	416	415
3·0	413	412	411	409	408	406	405	404	403	401
3·1	400	399	397	396	395	394	392	391	390	389
3·2	387	386	385	384	383	381	380	379	378	377
3·3	376	375	373	372	371	370	369	368	367	366
3·4	365	364	363	361	360	359	358	357	356	355
3·5	354	353	352	351	350	349	348	347	346	345
3·6	344	343	342	342	341	340	339	338	337	336
3·7	335	334	333	332	331	331	330	329	328	327
3·8	326	325	325	324	323	322	321	320	320	319
3·9	318	317	316	315	315	314	313	312	312	311
4·0	310	309	308	308	307	306	305	305	304	303
4·1	302	302	301	300	299	299	298	297	297	296
4·2	295	294	294	293	292	292	291	290	290	289
4·3	288	288	287	286	286	285	284	284	283	282
4·4	282	281	280	280	279	279	278	277	277	276
4·5	276	275	274	274	273	272	272	271	271	270
4·6	270	269	268	268	267	267	266	265	265	264
4·7	264	263	263	262	262	261	260	260	259	259
4·8	258	258	257	257	256	256	255	255	254	254
4·9	253	252	252	251	251	250	250	249	249	248
5·0	248	247	247	246	246	245	245	245	244	244
5·1	243	243	242	242	241	241	240	240	239	239
5·2	238	238	238	237	237	236	236	235	235	234
5·3	234	233	233	233	232	232	231	231	230	230
5·4	230	229	229	228	228	227	227	227	226	226
5·5	225	225	225	224	224	223	223	223	222	222
5·6	221	221	221	220	220	219	219	219	218	218
5·7	218	217	217	216	216	216	215	215	214	214
5·8	214	213	213	213	212	212	212	211	211	210
5·9	210	210	209	209	209	208	208	208	207	207

This table gives conversions from photon energy measured in electron volts (eV) to wavelength in nanometres, over the range likely to be of use to fluorescence microscopists. The table was calculated by computer according to the formula $E = 1239\cdot77/\lambda$. Some of the numbers were checked against independent calculations on a different machine with a different program. Because of the reciprocal nature of the numbers, the table can be used on a vice versa basis, with appropriate changes of decimal point as required.

Appendix 2

Standard fluorescein solution for microfluorometry

(Jongsma, Hijmans & Ploem, 1971)

Solution A: Take a round-bottomed flask of 50 ml capacity; add 10 ml of freshly prepared 5% sodium hydroxide in 96% ethanol; then add 20·8 mg of fluorescein diacetate. Warm the flask under reflux in a water-bath and swirl gently until the fluorescein diacetate is completely hydrolysed and dissolved (about 10 min). Allow the mixture to cool to room temperature. Add 20 ml of distilled water. Adjust to pH 8·5 with 1 M HCl added dropwise (about 12 ml will be required). Transfer the contents of the flask completely to a 50 ml volumetric flask and make up the volume to 50 ml with distilled water.

Solution B: In a 150 ml beaker, place: 1·21 g of trishydroxymethyl aminomethane (Tris; toxic!), 82·0 g of glycerol, 2 ml of 1 M HCl, and 10 ml of distilled water. Stir to dissolve the Tris (about 30 min). Adjust to pH 8·5 (using electrode suitable for Tris buffers) with 1 M HCl added dropwise.

Standard solution: Transfer Solution B to a 100 ml volumetric flask; add 5 ml of Solution A, mix, and make up the volume to 100 ml with distilled water. Store in a glass-stoppered dark bottle at 4 °C.

References

These references are arranged according to current Anglo-American cataloguing rules (Gorman & Winkler, 1978). For names beginning with van, van der, etc., or with von, see under the capitalized surname; for names beginning with Van, see under Van. Names beginning with de appear under de.

Agnati, L.F., Andersson, K., Wiesel, F. & Fuxe, K. (1979) A method to determine dopamine levels and turnover rate in discrete dopamine nerve terminal systems by quantitative use of dopamine fluorescence obtained by Falck–Hillarp methodology. *Journal of Neuroscience Methods* **1**, 365–373.

Agrup, G., Björklund, A., Falck, B., Jacobsson, S., Lindvall, O., Rorsman, H. & Rosengren, E. (1977) Fluorescence histochemical demonstration of dopa thioethers by condensation with gaseous formaldehyde. *Histochemistry* **52**, 179–186.

Alho, H. (1984) Microfluorometric quantitation of catecholamine turnover in the sympathetic neurons of rat. *Histochemistry* **80**, 363–366.

Alho, H., Partanen, M. & Hervonen, A. (1983) Microfluorimetric quantification of catecholamine fluorescence in rat sympathetic ganglia. *Histochemical Journal* **15**, 1203–1215.

Alho, H., Takala, T. & Hervonen, A. (1980) Microspectrofluorimetric quantitation of catecholamine fluorescence in rat SIF cells and paraganglia: effect of hypoxia. In: Eränkö, O., Soinila, S. & Päivärinta, H. (Eds.) *Histochemistry and cell biology of autonomic neurons, SIF cells and paraneurons*, pp. 215–218. Raven Press, New York.

Allen, R.D. & Allen, N.S. (1983) Video-enhanced microscopy with a computer frame memory. *Journal of Microscopy* **129**, 3–17.

Amos, W.B. (1988) Results obtained with a sensitive confocal scanning system designed for epifluorescence. *Cell Motility and the Cytoskeleton* **10**, 54–61.

Anders, J.J. & Salopek, M. (1989) Meningeal cells increase *in vitro* astrocytic gap junctional communication as measured by fluorescence recovery after laser photobleaching. *Journal of Neurology* **18**, 257–264.

Andersson, K. & Eneroth, P. (1985) The effects of acute and chronic treatment with triiodothyronine and thyroxine on the hypothalamic and telencephalic catecholamine nerve terminal systems of the hypophysectomized male rat. Chronic treatment modulates catecholamine utilization in discrete catecholamine nerve terminal systems. *Neuroendocrinology* **40**, 398–408.

Andersson, K., Nilsen, O.G., Toftgård, R., Eneroth, P., Gustaffson, J.-Å., Battistini, N. & Agnati, L.F. (1983) Increased turnover in several hypothalamic noradrenaline nerve terminal systems and changes in prolactin secretion in the male rat by exposure to various concentrations of toluene. *Neurotoxicology* **4**, 43–56.

Andreoni, A., Cova, S., Bottiroli, G. & Prenna, G. (1979) Fluorescence of complexes of quinacrine mustard with DNA. II. Dependence on the staining conditions. *Photochemistry and Photobiology* **29**, 951–958.

Andreoni, A., Longoni, A., Sacchi, C.A. & Svelto, O. (1980) Laser fluorescent microirradiation. A new technique. In: Goldman, L. (Ed.) *The biomedical laser: technology and applications*, pp. 69–83. Springer-Verlag, Berlin, Heidelberg & New York.

Andreoni, A., Sacchi, C.A., Bottiroli, G. & Prenna, G. (1975) Pulsed tunable laser in cytofluorometry: a study of the fluorescence pattern of chromosomes. In: Joussot-Dubien, J. (Ed.) *Lasers in physical chemistry and biophysics*, pp. 413–424. Elsevier Scientific, Amsterdam & New York.

Andrews, D.L. (1986) *Lasers in chemistry*. 160 pp. Springer-Verlag, Berlin, Heidelberg & New York.

Anon (1971) *The colour index*. 3rd edn. The Society of Dyers & Colourists, Bradford, UK.

Ansley, H.R. & Orstein, L. (1971) Enzyme histochemistry and differential white cell counts on the Technicon Hemalog D. *Advances in Automated Analysis* **1**, 437–443.

Aoki, K. (1981) Studies on the fluorescence of zircons. I. Types of fluorescence spectrum. *Bulletin of the Yamagata University (Natural Science)* **10**, 219–226.

Aoki, K. (1982a) Microspectrofluorometry of colorless zircons. *Bulletin of the Yamagata University (Natural Science)* **10**, 325–333.

Aoki, K. (1982b) A new method of discerning zircons by fluorescence spectrum. *Naturwissenschaften* **69**, 184.

Araki, R. & Yamada, M.-O. (1986) Measurements of multiple parameters in fluorometry: fluorescence intensity, lifetime and anisotropy. *Acta histochemica et cytochemica* **19**, 83–93.

Argauer, R.J. & White, C.E. (1964) Fluorescent compounds for calibration of excitation and emission units of spectrofluorometer. *Analytical Chemistry* **36**, 368–371.

Armitage, P. (1971) *Statistical methods in medical research*. 504 pp. Blackwell Scientific Publications, Oxford & Edinburgh.

Arndt-Jovin, D.J., Latt, S.A., Striker, G. & Jovin, T.M. (1979) Fluorescence decay analysis in solution and in a microscope of DNA and chromosomes stained with quinacrine. *Journal of Histochemistry and Cytochemistry* **27**, 87–95.

Arndt-Jovin, D.J., Robert-Nicoud, M., Baurschmidt, P. & Jovin, T.M. (1985) Immunofluorescence localization of Z-DNA in chromosomes: quantitation by scanning microphotometry and computer-assisted image analysis. *Journal of Cell Biology* **101**, 1422–1433.

Arndt-Jovin, D.J., Robert-Nicoud, M. & Jovin, T.M. (1990) Probing DNA structure and function with a multi-wavelength fluorescence confocal laser microscope. *Journal of Microscopy* **157**, 61–72.

Ash, E.A. (Ed.) (1980) *Scanned image microscopy*. 461 pp. Academic Press, London & New York.

Aubin, J.E. (1979) Autofluorescence of viable cultured mammalian cells. *Journal of Histochemistry and Cytochemistry* **27**, 36–43.

Axelrod, D. (1985) Fluorescence photobleaching techniques and lateral diffusion. In: Bayley, P.M. & Dale, R.E. (Eds.) *Spectroscopy and the dynamics of molecular biological systems*, pp. 163–176. Academic Press, London.

Axelrod, D., Koppel, D., Schlesinger, E., Elson, E. & Webb, W. (1976) Mobility measurement by analysis of fluorescence photobleaching recovery kinetics. *Biophysical Journal* **16**, 1005–1069.

Axelsson, S., Björklund, A., Falck, B., Lindvall, O. & Svensson, L.Å. (1973) Glyoxylic acid condensation: a new fluorescence method for the histochemical demonstration of biogenic amines. *Acta physiologica scandinavica* **87**, 57–62.

Bacopoulos, N.G., Bhatnagar, R.K., Schute, W.J. & Van Orden, L.S. (1975) On the use of the fluorescence histochemical method to estimate catecholamine content in brain. *Neuropharmacology* **14**, 291–299.

Baer, H.U., Baer-Suryadinata, C., Segantini, P. & Bollinger, A. (1985) Capillary damage after frostbite evaluated by fluorescence videomicroscopy. *Schweizerische Medizinische Wochenschrift* **115**, 479–483.

Bahr, G.F. & Wied, G.L. (1966) Cytochemical determination of DNA and basic protein in bull spermatozoa. Ultraviolet spectrophotometry, cytophotometry, and microfluorometry. *Acta cytologica* **10**, 393–412.

Balaban, R.S., Kurtz, I., Cascio, H.E. & Smith, P.D. (1986) Microscopic spectral imaging using a video camera. *Journal of Microscopy* **141**, 31–39.

Ballard, D.H. & Brown, C.M. (1982) *Computer vision*. Prentice-Hall, New York.

Banga, I. & Bihardi-Varga, M. (1974) Investigations of free and elastin-bound fluorescent substances present in the atherosclerotic lipid and calcium-plaques. *Connective Tissue Research* **2**, 237–241.

Barden, H., Aviles, F. & Rivers, W. (1979) Interference filter microfluorometry and quantitation of fluorescence of neuromelanin and lipofuscin age pigments in human brain. *Pigment Cell* **4**, 263–269.

Barni, S., Gerzeli, G. & Novelli, G. (1984) Cytochemistry of cell surface sialoglycoconjugates in endometrial adenocarcinoma. Effects of medroxyprogesterone therapy. *Applied Pathology* **2**, 135–145.

Baroni, B. (1933) Contributo allo studio dei melanomi cutanei al lume di un moderno mezzo d'indagine: del microscopio a fluorescenza. *Archivo Italiano di Dermatologia* **9**, 543–586.

Barrows, G.H., Sisken, J.E., Allegra, J.C. & Grasch, S.D. (1984) Measurement of fluorescence using digital integration of video images. *Journal of Histochemistry and Cytochemistry* **32**, 741–746.

Bastos, A.L., Marques, D., da Silva, J. F., Nunes, J.B.M., Correia, A.D., Vigário, J.D. & Terrinha, A.M. (1968) Primary inducible fluorescence in secretory granules of tumour cells stained by a Toluidine Blue dye. *Zeitschrift für Naturforschung* **23b**, 969–975.

Bauman, J.G.J., Bayer, J.A. & Dekken, H. van (1990) Fluorescent *in-situ* hybridization to detect cellular RNA by flow cytometry and confocal microscopy. *Journal of Microscopy* **157**, 73–82.

Bauman, J.G.J., Ploeg, M. van der & Duijn, P. van (1984) Fluorescent hybridocytochemical procedures: DNA/RNA hybridization in situ. In: Chayen, J. & Bitensky, L. (Eds.) *Investigative microtechniques in medicine and biology*, pp. 41–87. Marcel Dekker, New York & Basel.

Beddard, G.S. & West, M.A. (Eds.) (1981) *Fluorescent probes*. 235 pp. Academic Press, London.

Benedetti, P.A. & Lenci, F. (1977) In vivo microspectrofluorometry of photoreceptor pigments in *Euglena gracilis*. *Photochemistry and Photobiology* **26**, 315–318.

Benson, D.M., Bryan, J., Plant, A.L., Gotto, A.M. & Smith, L.C. (1985) Digital imaging fluorescence microscopy: spatial heterogeneity of photobleaching rate constants in individual cells. *Journal of Cell Biology* **100**, 1309–1323.

Benson, D.M. & Knopp, J.A. (1984) Effect of tissue absorption and microscope optical parameters on the depth of penetration for fluorescence and reflectance measurements of tissue samples. *Photochemistry and Photobiology* **39**, 495–502.

Benson, D.M., Knopp, J.A. & Longmuir, I.S. (1980) Intracellular oxygen measurements of mouse liver cells using quantitative fluorescence video microscopy. *Biochimica et biophysica acta* **591**, 187–197.

Bergquist, N.R. (1973) The pulsed dye laser as a light source for the fluorescent antibody technique. *Scandinavian Journal of Immunology* **2**, 37–44.

Berlin, R.D. & Oliver, J.M. (1980) Surface kinetics during mitosis. II. Quantitation of pinocytosis and kinetic characterization of the mitotic cycle with a new fluorescence technique. *Journal of Cell Biology* **85**, 660–671.

Berns, M.W. (1974) *Biological microirradiation*. 152 pp. Prentice-Hall, Englewood Cliffs, NJ.

Berns, M.W. (1979) Fluorescence analysis of cells using a laser light source. *Cell Biophysics* **1**, 1–13.

Berns, M.W. & Salet, C. (1972) Laser microbeams for partial cell irradiation. *International Review of Cytology* **33**, 131–155.

Berns, M.W., Siemens, A.E. & Walter, R.J. (1984) Mitochondrial fluorescence patterns in Rhodamine 6G-stained myocardial cells in vitro. Analysis by real-time computer video microscopy and laser microspot excitation. *Cell Biophysics* **6**, 263–277.

Bertero, M., Boccacci, P., Brakenhoff, G.J., Malfanti, F. & Voort, H.T.M. van der (1990) Three-dimensional image restoration and super-resolution in fluorescence confocal microscopy. *Journal of Microscopy* **157**, 3–20.

Björklund, A., Ehinger, B. & Falck, B. (1968a) A possibility for differentiating dopamine from noradrenaline in tissue sections by microspectrofluorometry. *Acta physiologica scandinavica* **72**, 253–254.

Björklund, A., Ehinger, B. & Falck, B. (1968b) A method for differentiating dopamine from noradrenaline in tissue sections by microspectrofluorometry. *Journal of Histochemistry and Cytochemistry* **16**, 263–270.

Björklund, A. & Falck, B. (1973) Cytofluorometry of biogenic amines in the Falck–Hillarp method. Structural identification by spectal analysis. In: Thaer, A.A. & Sernetz, M. (Eds.) *Fluorescence techniques in cell biology*, pp. 171–181. Springer-Verlag, Berlin, Heidelberg & New York.

Björklund, A., Falck, B. & Håkanson, R. (1968) Histochemical demonstration of tryptamine. Properties of the formaldehyde-induced fluorophores of tryptamine and related indole compounds in models. *Acta physiologica scandinavica* Supplementum **318**.

Björklund, A., Falck, B. & Lindvall, O. (1975) Microspectrofluorometric analysis of cellular monoamines after formaldehyde or glyoxylic acid condensation. In: Bradley, P.B. (Ed.) *Methods in brain research*, pp. 249–294. John Wiley & Sons, London.

Björklund, A., Falck, B. & Owman, C. (1972) Fluorescence microscopic and microspectrofluorometric techniques for the cellular localization and characterization of biogenic amines. In: Rall, J.E. & Kopin, I.J. (Eds.) *Methods of investigative and diagnostic endocrinology*, vol. **1**, pp. 318–368. North-Holland, Amsterdam.

Björklund, A., Håkanson, R., Lindvall, O. & Sundler, F. (1973) Fluorescence histochemical demonstration of peptides with NH_2 or COOH-terminal tryptophan or dopa by condensation with glyoxylic acid. *Journal of Histochemistry and Cytochemistry* **21**, 253–265.

Björklund, A., Nobin, A. & Stenevi, U. (1971) Acid catalysis of the formaldehyde condensation reaction for a sensitive histochemical demonstration of tryptamines and 3-methoxylated phenylethylamines. 2. Characterization of the amine fluorophores and application to tissues. *Journal of Histochemistry and Cytochemistry* **19**, 286–298.

Björklund, A. & Stenevi, U. (1970) Acid catalysis of the formaldehyde condensation reaction for a sensitive histochemical demonstration of tryptamines and 3-methoxylated phenylethylamines. *Journal of Histochemistry and Cytochemistry* **18**, 794–802.

Björn, L.O. & Björn, G.S. (1986) Studies on energy dissipation in phycobilisomes using the Kennard–Stepanov relation between absorption and fluorescence emission spectra. *Photochemistry and Photobiology* **44**, 535–542.

Blanchard, S.G., Chang, K.J. & Cuatrecasas, P. (1983) Visualization of enkephalin receptors by image-intensified fluorescence microscopy. In: Conn, P.M. (Ed.) *Hormone action*, Part H: *Neuroendocrine peptides. Methods in Enzymology* **103**, Part H. Academic Press, Orlando, FL.

Blazak, W.F., Overstreet, J.W., Katz, D.F. & Hanson, F.W. (1982) A competitive *in vitro* assay of human sperm fertilizing ability utilizing contrasting fluorescent sperm markers. *Journal of Andrology* **3**, 165–171.

Böck, G., Hilchenbach, M., Schauenstein, K. & Wick, G. (1985) Photometric analysis of antifading reagents for immunofluorescence with lasr and conventional illumination sources. *Journal of Histochemistry and Cytochemistry* **33**, 699–705.

Boender, J. (1984) Fluorescein-diacetate, a fluorescent dye compound stain for rapid evaluation of the viability of mammalian oocytes prior to in vitro studies. *Veterinary Quarterly* **6**, 236–240.

Böhm, N. & Sprenger, E. (1968) Fluorescence cytophotometry: a valuable method for the quantitative determination of nuclear Feulgen-DNA. *Histochemie* **16**, 100–118.

Böhm, N. & Sprenger, E. (1970) 'Nine-position test' for proof of even illumination of the light field in fluorescence microspectrophotometry. *Histochemie* **22**, 59–61.

Böhm, N., Sprenger, E. & Sandritter, W. (1970) Die Feulgen-Reaktion mit Acriflavin-SO_2. Ein Vergleich zwischen Absorptions- und Fluoreszenz Cytophotometrie anhand von Hydrolysekurven. *Acta histochemica* **35**, 324–329.

Böhm, N., Sprenger, E. & Sandritter, W. (1973) Absorbance and fluorescence cytophotometry of nuclear Feulgen DNA. A comparative study. In: Thaer, A.A. & Sernetz, M. (Eds.) *Fluorescence techniques in cell biology*, pp. 67–77. Springer-Verlag, Berlin, Heidelberg & New York.

Boldt, M., Harbig, K., Weidemann, G. & Lübbers, D.W. (1980) A sensitive dual wavelength microspectrophotometer for the measurement of tissue fluorescence and reflectance. *Pflügers Archiv* **385**, 167–173.

Bollinger, A. (1988) Fluorescence videomicroscopy. *Clinical Haematology* **8**, 379–384.

Bollinger, A., Franzeck, U.K., Isenring, G., Frey, J. & Jäger, K. (1983) Möglichkeiten der Fluoreszenz-Videomikroskopie in der Gefässdiagnostik. *Deutsche medizinische Wochenschrift* **108**, 422–424.

Bollinger, A., Franzeck, U.K. & Jäger, K. (1983) Quantitative capillaroscopy in man using fluorescence video-microscopy. *Progress in Applied Microcirculation* **3**, 97–118.

Bollinger, A., Frey, J., Jäger, K., Furrer, J., Seglias, J. & Siegenthaler, W. (1982) Patterns of diffusion through skin capillaries in patients with long-term diabetes. *New England Journal of Medicine* **307**, 1305–1310.

Bollinger, A., Jäger, Rotem, A., Timeus, C. & Mahler, F. (1979) Diffusion, pericapillary distribution and clearance of Na-fluorescein in the human nailfold. *Pflügers Archiv* **382**, 137.

Bonner, W.A., Hulett, H.R., Sweet, R.G. & Herzenberg, L.A. (1972) Fluorescence activated cell sorting. *Reviews of Scientific Instrumentation* **43**, 404–409.

Børessen, H.C. & Parker, C.A. (1966) Some precautions required in the calibration of fluorescence spectrometers in the ultraviolet region. *Analytical Chemistry* **38**, 1073–1074.

Borgström, E. & Wahren, B. (1985) Quantitative analysis of the A, B and H isoantigens in single transitional carcinoma cells. *Journal of Urology* **134**, 199–202.

Borkman, R.F., Tassin, J.D. & Lerman, S. (1981) Fluorescence lifetimes of chromophores in intact human lenses and lens proteins. *Experimental Eye Research* **32**, 313–322.

Borst, M. & Königsdorfer, H. (1929) *Untersuchungen über die Porphyrie mit besonderer Berücksichtigung der Porphyria congenita*. Hirzel-Verlag, Leipsig.

Borst, W.L., Hamid, H.A. & Crelling, J.C. (1985) Laser induced fluorescence microscopy. Proceedings of 1985 *International conference on coal science*, 28–31 October 1985, Sydney, NSW, pp. 649–652. Pergamon Press, Sydney & New York.

Bosshard, U. (1964) Fluoreszenzmikroskopische Messung des DNS-Gehaltes von Zellkernen. *Zeitschrift für Wissenschaftliche Mikroskopie* **65**, 391–408.

Bottiroli, G., Cionini, P.G., Docchio, F. & Sacchi, C.A. (1984) *In situ* evaluation of the functional state of chromatin by means of quinacrine mustard staining and time resolved fluorescence microscopy. *Histochemical Journal* **16**, 223–233.

Bottiroli, G., Prenna, G., Andreoni, A., Sacchi, C.A. & Svelto, O. (1979) Fluorescence of complexes of quinacrine mustard with DNA. I. Influence of the DNA base composition on the decay time in bacteria. *Photochemistry and Photobiology* **29**, 23–28.

Boyde, A. (1985) The tandem scanning reflected light microscope. Part 2 – Pre-Micro '84 applications at UCL. *Proceedings of the Royal Microscopical Society* **20**, 131–139.

Boyde, A. (1987) Applications of tandem scanning reflected light microscopy and three-dimensional imaging. *Annals of the New York Academy of Sciences* **483**, 428–439.

Boyde, A., Jones, S.J., Taylor, M.L., Wolfe, L.A. & Watson, T.F. (1990) Fluorescence in the tandem scanning microscope. *Journal of Microscopy* **157**, 39–50.

Boyde, A. & Watson, T.F. (1989) Fluorescence mode in the tandem scanning microscope. *Proceedings of the Royal Microscopical Society* **24**, 1.

Brakenhoff, G.J. (1979) Imaging modes in confocal scanning light microscopy. *Journal of Microscopy* **117**, 233–242.

Brakenhoff, G.J., Blom, P. & Barends, P. (1979) Confocal scanning light microscopy with high aperture immersion lenses. *Journal of Microscopy* **117**, 219–232.

Brakenhoff, G.J., Voort, H.T.M. van der, Spronsen, E.A. & Nanninga, N. (1986) Three-dimensional imaging by confocal scanning microscopy. *Annals of the New York Academy of Sciences* **483**, 405–415.

Brand, L. & Gohlke, J.R. (1972) Fluorescent probes for structure. *Annual Review of Biochemistry* **41**, 843–868.

Bright, G.R. & Taylor, D.L. (1986) Imaging at low light level in fluorescence microscopy. In: Taylor, D.L., Waggoner, A.S., Lanni, F., Murphy, R.F. & Birge, R.R. (Eds.) *Applications of fluorescence in the biomedical sciences*, pp. 257–288. Alan R. Liss, New York.

Broekaert, D., Ostveldt, P. van, Coucke, P., de Bersaques, J., Gillis, E. & Reyniers, P. (1986) Nuclear differentiation and ultimate fate during epidermal keratinization. *Archives of Dermatological Research* **279**, 100–111.

Brünger, A., Peters, R. & Schulten, K. (1985) Continuous fluorescence microphotolysis to observe lateral diffusion in membranes – theoretical methods – applications. *Journal of Chemical Physics* **82**, 2147–2160.

Burstone, M.S. (1958) The relationship between fixation and techniques for the histochemical localization of hydrolytic enzymes. *Journal of Histochemistry and Cytochemistry* **6**, 322–339.

Burton, J.L. & Bank, H.L. (1986) Analog enhancement of videomicroscope images. *Journal of Microscopy* **142**, 301–309.

Bussolati, G., Rost, F.W.D. & Pearse, A.G.E. (1969) Fluorescence metachromasia in polypeptide hormone-producing cells of the APUD series, and its signficance in relation to the structure of the precursor protein. *Histochemical Journal* **1**, 517–530.

Byrkit, D. (1987) *Statistics today, a comprehensive introduction.* 850 pp. Benjamin/Cummings, Menlo Park, CA.

Campbell, R.C. (1989) *Statistics for biologists*, 3rd edn. 446 pp. Cambridge University Press, Cambridge.

Candy, B.H. (1985a) Photomultiplier characteristics and practice relevant to photon counting. *Review of Scientific Instruments* **56**, 183–193.

Candy, B.H. (1985b) Photon counting circuits. *Review of Scientific Instruments* **56**, 194–200.

Carlsson, K. (1990) Scanning and detection techniques used in a confocal scanning laser microscope. *Journal of Microscopy* **157**, 21–28.

Caspersson, T., Farber, S., Foley, G.E., Kudynowski, J., Modest, E.J., Simonsson, W., Wagh, W. & Zech, L. (1968) Chemical differentiation along metaphase chromosomes. *Experimental Cell Research* **49**, 219–222.

Caspersson, T., Hillarp, N. Å. & Ritzén, M. (1966) Fluorescence microspectrofluorometry of cellular catecholamines and 5-hydroxytryptamine. *Experimental Cell Research* **42**, 415–428.

Caspersson, T., Jacobsson, F., Lomakka, G., Svensson, G. & Säfström, R. (1953) A high resolution ultramicrospectrophotometer for large-scale biological work. *Experimental Cell Research* **5**, 560–563.

Caspersson, T. & Lomakka, G. (1970) Recent progress in quantitative cytochemistry: instrumentation and results. In: Wied, G.L. & Bahr, G.F. (Eds.) *Introduction to quantitative cytochemistry II*, pp. 27–56. Academic Press, New York & London.

Caspersson, T., Lomakka, G. & Rigler R. (1965) Registierender Fluoreszenz-mikrospektrograph zur Bestimmung der Primär- und Secundärfluoreszenz verschiedener Zellsubstanzen. *Acta histochemica* Supplementum **6**, 123–126.

Caspersson, T., Lomakka, G. & Zech, L. (1971) The 24 fluorescence patterns of the human metaphase chromosomes – distinguishing characteristics and variability. *Hereditas* **67**, 89–102.

Cegrell, L., Falck, B. & Rosengren, A.-M. (1970) Extraction of dopa from the integument of pigmented animals. *Acta physiologica scandinavica* **78**, 65–69.

Chance, B. (1962) Kinetics of enzyme reactions within single cells. *Annals of the New York Academy of Sciences* **97**, 431–448.

Chance, B., Jamieson, D. & Williamson, J.R. (1966) Control of the oxidation–reduction state of reduced pyridine nucleotides in vivo and in vitro by hyperbaric oxygen. In: Brown, I.W. & Cox, B.G. (Eds.) *Proceedings of the Third International Conference on Hyperbaric Medicine*, pp. 15–41. National Academy of Sciences, Washington, DC.

Chance, B. & Legallais, V. (1959) Differential microfluorimeter for the localization of reduced pyridine nucleotide in living cells. *Review of Scientific Instruments* **30**, 732–735.

Chance, B., Mela, L. & Wong, D. (1968) Flavoproteins of the respiratory chain. In: Yagi, K. (Ed.) *Flavins and flavoproteins*, p. 107. Baltimore University Park Press, Baltimore; University of Tokyo Press, Tokyo.

Chance, B. & Thorell, B. (1959a) Localization and kinetics of reduced pyridine nucleotide in living cells by microfluorometry. *Journal of Biological Chemistry* **234**, 3044–3050.

Chance, B. & Thorell, B. (1959b) Fluorescence measurements of mitochondrial pyridine nucleotide in aerobiosis and anaerobiosis. *Nature* **184**, 931–934.

Chance, B., Williamson, J.R., Jamieson, D. & Schoener, B. (1965) Properties and kinetics of reduced pyridine nucleotide fluorescence of the isolated and *in vitro* heart. *Biochemische Zeitschrift* **341**, 357–377.

Changaris, D.G., Combs, J. & Severs, W.B. (1977) A microfluorescent PAS method for the quantitative demonstration of cytoplasmic 1,2-glycols. *Histochemistry* **52**, 1–5.

Chapman, J.H., Förster, T., Kortüm, G., Parker, C.A., Lippert, E., Melhuish, W.H. & Nebbia, G. (1963) Proposal for standardization of methods of reporting fluorescence emission spectra. *Applied Spectroscopy* **17**, 171. Reprinted in: Udenfriend, S. (1969) *Fluorescence assay in biology and medicine*, pp. 592–593. Academic Press, New York.

Chen, R.F. (1967a) Some characteristics of the fluorescence of quinine. *Analytical Biochemistry* **19**, 374–387.

Chen, R.F. (1967b) Practical aspects of the calibration and use of the Aminco-Bowman spectrophoto-fluorometer. *Analytical Biochemistry* **20**, 339–357.

Chen, R.F. (1973) Extrinsic and intrinsic fluorescence of proteins. In: Guilbault, G.G. (Ed.) *Practical fluorescence: theory, methods and techniques*, pp. 467–541. Marcel Dekker, New York.

Chen, R.F. (1974) Fluorescence lifetime reference standard for the range 0·189 to 115 nanoseconds. *Analytical Biochemistry* **57**, 593–604.

Chen, R.F. & Scott, C.H. (1985) Atlas of fluorescence spectra and lifetimes of dyes attached to protein. *Analytical Letters* **18**, 393–421.

Chen, R.F., Vurek, G.G. & Alexander, N. (1967) Fluorescence decay times: proteins, coenzymes and other compounds. *Science* **156**, 949–951.

Cheng, S.C., Crozier, P.A. & Egerton, R.F. (1987) Deadtime corrections for a pulse counting system in electron energy loss spectroscopy. *Journal of Microscopy* **148**, 285–288.

Číhalíková, J., Doležel, J. & Novák, F.J. (1985) Cytofluorometric determination of nuclear DNA in plant cells using Auramine O. *Acta histochemica* **76**, 151–156.

Cohan, C.S., Hadley, R.D. & Kater, S.B. (1983) 'Zap axotomy': localized fluorescent excitation of single dye-filled neurons induces growth by selective axotomy. *Brain Research* **270**, 93–101.

Coleman, A.W. (1984) The fate of chloroplast DNA during cell fusion, zygote maturation and zygote germination in *Chlamydomonas reinhardii* as revealed by DAPI staining. *Experimental Cell Research* **152**, 528–540.

Coleman, A.W., Maguire, M.J. & Coleman, J.R. (1981) Mithramycin- and 4',6-Diamidino-2-phenylin-dole (DAPI)-DNA staining for fluorescence microspectrophotometric measurement of DNA in nuclei, plastids, and virus particles. *Journal of Histochemistry and Cytochemistry* **29**, 959–968.

Collins, V.P. & Thaw, H.H. (1983) The measurement of lipid peroxidations products (lipofuscin) in individual cultured human glial cells. *Mechanisms of Aging and Development* (Lausanne) **23**, 199–214.

Combs, J.W. (1973) An automatic microspectrophotometric system for rapid quantitative cytofluori-metry and reflectance autoradiography. *The Microscope* **21**, 11–21.

Corrodi, H. & Hillarp, N.Å. (1963) Fluoreszenzmethoden zur histochemischen Sichtbarmachung von Monoaminen. I. Identifizierung der fluoreszierenden Produktes aus Modellversuchen mit 6,7-Dimeth-oxyisochinolinderivaten und Formaldehyd. *Helvetia chimica acta* **46**, 2425–2430.

Corrodi, H. & Jonsson, G. (1965) Fluorescence methods for the histochemical demonstration of monoamines: 4. Histochemical differentiation between dopamine and noradrenaline in models. *Journal of Histochemistry and Cytochemistry* **13**, 484–487.

Corrodi, H. & Jonsson, G. (1966) Fluoreszenzmethoden zur histochemischen Sichtbarmachung von Monoaminen. 6. Identifizierung der fluoreszierenden Produkte aus *m*-Hydroxyphenyläthylaminen und Formaldehyd. *Helvetia chimica acta* **49**, 798–806.

Corrodi, H. & Jonsson, G. (1967) The formaldehyde fluorescence method for the histochemical demonstration of biogenic amines. A review on the methodology. *Journal of Histochemistry and Cytochemistry* **15**, 65–78.

Coulter, W.H. (1949) Means for counting particles suspended in a fluid. US Patent No. 2 656 508. Filed 27 August 1949, issued 20 October 1953.

Cova, S., Longoni, A., Giordano, P.A. & Freitas, I. (1986) Minimization of photofading and drift-induced errors by spectrum scanning strategy. *Analytical Chemistry* **48**, 3148–3153.

Cova, S., Prenna, G. & Mazzini, G. (1972) Microspectrofluorometry by single photon multiscaling. *Proceedings of the 4th International Congress of Histochemistry and Cytochemistry*, pp. 161–162. Kyoto.

Cova, S., Prenna, G. & Mazzini, G. (1974) Digital microspectrofluorometry by multichannel scaling and single photon detection. *Histochemical Journal* **6**, 279–299.

Cowden, R.R. & Curtis, S.K. (1973) Fluorescence cytochemical studies of chromosomes: quantitative applications of fluorescein mercuric acetate. In: Thaer, A.A. & Sernetz, M. (Eds.) *Fluorescence techniques in cell biology*, pp. 135–149. Springer-Verlag, Berlin, Heidelberg & New York.

Cowden, R.R. & Curtis, S.K. (1981) Microfluorometric investigations of chromatin structure. II. Mordant fluorochroming with ions that complex with morin. *Histochemistry* **72**, 391–400.

Cowell, J.K. & Franks, L.M. (1980) A rapid method for accurate DNA measurements in single cells in situ using a simple microfluorometer and Hoechst 3328 as a quantitative fluorochrome. *Journal of Histochemistry and Cytochemistry* **28**, 206–210.

Cowen, T. & Burnstock, G. (1982) Image analysis of catecholamine fluorescence. *Brain Research Bulletin* **9**, 81–86.

Cox, B.A., Yielding, L.W. & Yielding, K.L. (1984) Subcellular localization of photoreactive ethidium analogs in *Trypanosoma brucei* by fluorescence microscopy. *Journal of Parasitology* **70**, 694–702.

Cox, I.J. (1984) Scanning optical fluorescence microscopy. *Journal of Microscopy* **133**, 149–154.

Cox, I.J., Sheppard, C.J.R. & Wilson, T. (1982) Super-resolution by confocal fluorescence microscopy. *Optik* **60**, 391–396.

Crelling, J.C. & Bensley, D.F. (1980) Petrology of cutinite-rich coal from the Roaring Creek Mine, Parke County, Indiana. In: Langenheim, R.L. & Mann, C.J. (Eds.) *Middle and late Pennsylvanian strata on margin of Illinois basin, Vermilion County, Illinois, and Vermilion and Parke Counties, Indiana*, pp. 93–104. Tenth Annual Field Conference, Great Lakes Section of Society of Economic Paleontologists and Mineralogists.

Crelling, J.C. & Bensley, D.F. (1984) Characterization of coal macerals by fluorescence microscopy. In: Winans, R.E. & Crelling, J.C. (Eds.) *Chemistry and characterization of coal macerals*. 185th meeting of the American Chemical Society: Chemistry and characterization of coal macerals. Seattle, WA. 20–25 March 1983. *ACS Symposium Series* **252**, 33–45.

Crelling, J.C. & Dutcher, R.R. (1979) Secondary resinite in some Utah coals. *Geological Society of America Abstracts with Programs* (Annual Meeting) **11**, 406.

Crelling, J.C. & Dutcher, R.R. (1980) Fluorescent macerals in Colorado coking coals. In: Carter, L.M. (Ed.) *Proceedings of the fourth symposium on the geology of Rocky Mountain coal, 1980*. Colorado Geological Survey, Resource Series **10**, pp. 58–61.

Crelling, J.C., Dutcher, R.R. & Lange, R.V. (1982) Petrographic and fluorescence properties of resinite macerals from western U.S. coals. In: Gurgel, K.D. (Ed.) *Proceedings of the fifth symposium on the geology of Rocky Mountain coal – 1982*, pp. 187–191. Bulletin 118, Utah Geological and Mineral Survey.

Cripps, D.J., Hawgood, R.S. & Magnus, I.A. (1966). Iodide tungsten fluorescence microscopy for porphyrin fluorescence. *Archives of Dermatology* **93**, 129–134.

Crissman, H.A., Oka, M.S. & Steinkamp, J.A. (1976) Rapid staining methods for analysis of deoxyribo-nucleic acid and protein in mammalian cells. *Journal of Histochemistry and Cytochemistry* **24**, 64–71.

Crosland-Taylor, P.J. (1953) A device for counting small particles suspended in a fluid through a tube. *Nature* **171**, 37–38.

Cross, S.A.M., Ewen, S.W.B. & Rost, F.W.D. (1971) A study of methods for the cytochemical localization of histamine by fluorescence induced with *o*-phthalaldehyde or acetaldehyde. *Histochemical Journal* **3**, 471–476.

Cummins, J.T., Rahn, C.L. & Rahn, R.S. (1982) Microscopic observations on endogenous fluorochromes within a nerve fibre excited by a 325 nm He-Cd laser. *Journal of Microscopy* **127**, 277–285.

Curtis, S.K. & Cowden, R.R. (1983) Evaluation of five basic fluorochromes of potential use in microfluorometric studies of nucleic acids. *Histochemistry* **78**, 503–511.

Curtis, S.K. & Cowden, R.R. (1985) Microfluorometric estimates of proteins associated with murine hepatocyte and thymocyte nuclei, residual structures, and nuclear matrix derivatives. *Histochemistry* **82**, 331–339.

Daley, P.F., Raschke, K., Ball, J.T. & Berry, J.A. (1989) Topography of photosynthetic activity of leaves obtained from video images of chlorophyll fluorescence. *Plant Physiology* **90**, 1233–1238.

Daugherty, T.H. & Hjort, E.V. (1934) A new ultraviolet microscope illuminator. *Industrial and Engineering Chemistry, Analytical Edition* **6**, 370–371.

David, G.B. & Galbraith, W. (1975) The Denver universal microspectroradiometer (DUM). 1. General design and construction. *Journal of Microscopy* **103**, 135–178.

de Josselin de Jong, J.E., Jongkind, J.F. & Ywema, H.E. (1980) A scanning inverted microfluorometer with electronic shutter control for automatic measurements in micro-test plates. *Analytical Biochemistry* **102**, 120–125.

de la Torre, L. & Salisbury, G.W. (1962) Fading of Feulgen-stained bovine spermatozoa. *Journal of*

Histochemistry and Cytochemistry **10**, 39–41.

de Lerma, B. (1958) Die Anwendung von Fluoreszenzlicht in der Histochemie. *Handbuch der Histochemie*, vol. **1**, part 1, pp. 78–159. G. Fischer Verlag, Stuttgart.

de Permentier, P. (1981) Histochemical and biochemical studies of the kinetics of acid phosphatase in rat kidney. MSc thesis, University of New South Wales.

Dean, P.N. (1985) Methods of data analysis in flow cytometry. In: Van Dilla, M.A., Dean, P.N., Laerum, O.D. & Melamed, M.R. (Eds.) *Flow cytometry: instrumentation and data analysis*, pp. 195–221. Academic Press, London.

Dean, P.N. & Pinkel, D. (1978) High resolution dual laser flow cytometry. *Journal of Histochemistry and Cytochemistry* **26**, 622–627.

Deelder, A.M., Tanke, H.J. & Ploem, J.S. (1978) Automated quantitative immunofluorescence using the aperture defined microvolume (ADM) method. In: Knapp, W., Holubar, K. & Wick, G. (Eds.) *Immunofluorescence and related staining techniques*, pp. 31–44. Elsevier/North Holland, Amsterdam & New York.

Demko, P.C. & Todd, O.E. (1976) Magnetically stabilized xenon arc lamp. US Patent No. 3 988 626, filed 12 May 1975, granted 26 October 1976.

Derman, C. & Klein, M. (1959) *Probability and statistical inference for engineers*. 144 pp. Oxford University Press, Oxford.

Deyl, Z., Macek, K., Adam, M. & Vančiková, O. (1980) Studies on the chemical nature of elastin fluorescence. *Biochimica et biophysica acta* **625**, 248–254.

Dittrich, W. & Goehde, W. (1968) Automatic measuring and counting device for particles in a dispersion. British Patent No. 1 300 585. Filed 18 December 1968 in Germany, issued 20 December 1972.

Dittrich, W. & Goehde, W. (1969) Automatic measuring and counting device for particles in a dispersion. British Patent No. 1 305 923. Filed 18 April 1969 in Germany, published 7 February 1973.

Dixon, W.J. & Massey, F.J. (1968) *Introduction to statistical analysis*, 3rd edn. 638 pp. McGraw-Hill, New York.

Dobell, P., Cameron, A.R. & Kalreuth, W.D. (1984) Petrographic examination of low-rank coals from Saskatchewan and British Columbia, Canada, including reflected and fluorescent light microscopy, SEM, and laboratory oxidation procedures. *Canadian Journal of Earth Sciences* (*Journal Canadien des Sciences de la Terre*) **21**, 1209–1228.

Docchio, F., Ramponi, R., Sacchi, C.A., Bottiroli, G. & Freitas, I. (1982) Time resolved fluorescence microscopy of hematoporphryrin-derivative in cells. *Lasers in Surgery and Medicine* **2**, 21–28.

Docchio, F., Ramponi, R., Sacchi, C.A., Bottiroli, G. & Frietas, I. (1984) An automatic pulsed laser microfluorometer with high spatial and temporal resolution. *Journal of Microscopy* **134**, 151–160.

Docchio, F., Ramponi, R., Sacchi, C.A., Bottiroli, G. & Freitas, I. (1986) Time-resolved fluorescence microscopy: examples of applications to biology. *Analytical Chemistry* **58**, 85–100.

Dolbeare, F.A. & Smith, R.E. (1977) Flow cytometric measurements of peptidases with use of 5-nitrosalicylaldehyde and 4-methoxy-beta-naphthylamine derivatives. *Journal of Clinical Chemistry* **23**, 1485–1491.

Dolbeare, F.A. & Smith, R.E. (1979) Flow cytoenzymology: rapid enzyme analysis of single cells. In: Melamed, M.R., Mullaney, P.F. & Mendelsohn, M.L. (Eds.) *Flow cytometry and sorting*, pp. 317–334. John Wiley, New York.

Donáth, T. (1963) Fluoreszenzintensitätsmessungen an cytologischem und histologischem Material (methodischer Teil). *Mikroskopie* **18**, 7–13.

Donovan, R.M. & Goldstein, E. (1985) A charge coupled device-based image cytophotometry system for quantitative histochemistry and cytochemistry. *Journal of Histochemistry and Cytochemistry* **33**, 551–556.

dos Remedios, C.G., Millikan, R.G.C. & Morales, M.F. (1972) Polarization of tryptophan fluorescence from single striated muscle fibres. *Journal of General Physiology* **59**, 103–120.

dos Remedios, C.G., Yount, R.G. & Morales, M.F. (1972) Individual states in the cycle of muscle contraction. *Proceedings of the National Academy of Sciences, USA* **69**, 2542–2546.

Dowson, J.H. (1984) Partial recovery of fluorescence intensity after irradiation-induced fading of fluorescence, and its effects on techniques of quantitative fluorescence microscopy. *Experientia* **40**, 538–539.

Dowson, J.H. & Harris, S.J. (1981) Quantitative studies of the autofluorescence derived from neuronal lipofuscin. *Journal of Microscopy* **123**, 249–258.

Draaijer, A. & Houpt, P.M. (1988) A standard video-rate confocal laser-scanning reflection and fluorescence microscope. *Scanning* **10**, 139–145.

Drexhage, K.H. (1973) Structure and properties of laser dyes. In: Shäfer, F.P. (Ed.) *Dye lasers* (Topics in applied physics, vol. **1**), Chapter 4, pp. 144–193. Springer-Verlag, Berlin, Heidelberg & New York.

Duggan, J.X., DiCesare, J. & Williams, J.F. (1983) Investigations on the use of laser dyes as quantum counters for obtaining corrected fluorescence spectra in the near infrared. In: Eastwood, D. (Ed.) *New directions in molecular luminescence*, ASTM Special Technical Publication 822, American Society for Testing and Materials, pp. 112–126.

Duijn, P. van, Pascoes, E. & Ploeg, M. van der (1967) Theoretical and experimental aspects of enzyme determination in a cytochemical model system of polyacrylamide films containing alkaline phosphatase. *Journal of Histochemistry and Cytochemistry* **15**, 631–645.

Duijn, P. van & Ploeg, M. van der (1970) Potentialities of cellulose and polyacrylamide films as vehicles in quantitative cytochemical investigations on model substances. In: Wied, G.L. & Bahr, G.F. (Eds.) *Introduction to quantitative cytochemistry II*, pp. 223–263. Academic Press, New York.

Eaton, D.F. (Ed.) (1988) International Union of Pure and Applied Chemistry, Organic Chemistry Division, Commission on Photochemistry: Reference materials for fluorescence measurement. *Pure and Applied Chemistry* **60**, 1107–1114.

Eccles, M.J., Sim, M.E. & Tritton, K.P. (1983) *Low light level detectors in astronomy.* 187 pp. Cambridge University Press, Cambridge.

Eder, H. (1966) *Fluoreszenzcytophotometrische Untersuchungen an lebenden Lymphozyten.* Habilitationschrift, Giessen. (Cited in Sernetz, M. & Thaer, A. (1970) A capillary fluorescence standard for microfluorometry. *Journal of Microscopy* **91**, 43–52.)

Eder, H. (1986) Ein Fluorescenzphänomen am Nucleoprotein lebender Zellen unter Belichtung. *Acta histochemica* **79**, 67–81.

Eftink, M. (1983) Quenching-resolved emission anisotropy studies with single and multi-tryptophan-containing proteins. *Biophysical Journal* **43**, 323–334.

Ehinger, B., Falck, B., Persson, R. & Sporrong, B. (1968) Adrenergic and cholinesterase-containing neurons of the heart. *Histochemie* **16**, 197–205.

Einarsson, P., Hallman, H. & Jonsson, G. (1975) Quantitative microfluorimetry of formaldehyde induced fluorescence of dopamine in the caudate nucleus. *Medical Biology* **53**, 15–24.

Eisert, W.G. & Beisker, W. (1980) Epi-illumination optical design for fluorescence polarization measurements in flow systems. *Biophysical Journal* **31**, 97–112.

Ellinger, P. (1940) Fluorescence microscopy in biology. *Biological Review* **15**, 323–350.

Ellinger, P. & Hirt, A. (1929a) Mikroskopische Untersuchungen an lebenden Organen. I. Mitteilung: Methodik: Intravitalmikroskopie. *Zeitschrift für Anatomie und Entwicklungs-Geschichte* **90**, 791–802.

Ellinger, P. & Hirt, A. (1929b) Mikroskopische Untersuchungen an lebenden Organen. II. Mitteilung: zur Funktion der Froschniere. *Archiv für experimentelle Pathologie und Pharmakologie* **145**, 193–210.

Ellinger, P. & Hirt, A. (1930) Eine Methode zur Beobachtung lebender Organe mit stärksten Vergrösserung im Lumineszenzlicht (Intravitalmikroskopie). *Handbuch der biologischen Arbeitsmethode*, Abt. V, Teil 2, Heft 15, 1753–1764. Urban & Schwarzenberg, Berlin & Vienna.

Enerbäck, L. (1974) Berberine sulphate binding to mast cell polyanions: a cytofluorometric method for the quantitation of heparin. *Histochemistry* **42**, 301–313.

Enerbäck, L. & Gustavsson, G. (1977) Uptake of 5-hydroxytryptamine by mast cells in vivo. A cytometric study of mast cells and individual mast cell granules. *Histochemistry* **53**, 193–202.

Enerbäck, L. & Jarlstedt, J. (1975) A cytofluorometric and radiochemical analysis of the uptake and turnover of 5-hydroxytryptamine in mast cells. *Journal of Histochemistry and Cytochemistry* **23**, 128–135.

Enerbäck, L. & Johansson, K.-A. (1973) Fluorescence fading in quantitative fluorescence microscopy: a cytofluorometer for the automatic recording of fluorescence peaks of very short duration. *Histochemical Journal* **5**, 351–362.

Enerbäck, L., Kristensson, K. & Olsson, T. (1980) Cytophotometric quantification of retrograde axonal transport of a fluorescent tracer (primuline) in mouse facial neurons. *Brain Research* **186**, 21–32.

Eränkö, O. & Räisänen, L. (1961) Cytophotometric study on the chromaffin reaction, the iodate reaction and formalin-induced fluorescence as indicators of adrenaline and noradrenaline concentrations in the adrenal medulla. *Journal of Histochemistry and Cytochemistry* **9**, 54–58.

Es, A. van, Tanke, H.J., Baldwin, W.M., Oljans, P.J., Ploem, J.S. & Es, L.A. van (1983) Ratios of T lymphocyte subpopulations predict survival of cadaveric renal allografts in adult patients on low-dose corticosteroid therapy. *Clinical and Experimental Immunology* **52**, 13–20.

Ewen, S.W.B. & Rost, F.W.D. (1972) The histochemical demonstration of catecholamines by acid- and aldehyde-induced fluorescence: microspectrofluorometric characterization of the fluorophores in models. *Histochemical Journal* **4**, 59–69.

Fabiato, A. (1982) Fluorescence and differential light absorption recordings with calcium probes and potential-sensitive dyes in skinned cardiac cells. *Canadian Journal of Physiology and Pharmacology* **60**, 556–567.

Falck, B. (1962) Observations on the possibilities of the cellular localisation of monoamines by a fluorescence method. *Acta physiologica scandinavica* **56**, Supplementum **197**.

Falck, B. & Owman, C. (1965) A detailed methodological description of the fluorescence method for the cellular demonstration of biogenic amines. *Acta Universitatis Lundensis* Sectio II, No. 7, Lund (Sweden).

Feuerstein, I.A. & Kush, J. (1986) Blood platelet surface interactions on fibrinogen under flow as viewed with fluorescent video-microscopy. *Journal of Biomechanical Engineering* **108**, 49–53.

Fleming, G.R. & Beddard, G.S. (1978) CW mode-locked dye lasers for ultra fast spectroscopic studies. *Optics and Laser Technology* Oct., 257–264.

Fletcher, A.N. (1968) Fluorescence emission band shift with wavelength of excitation. *Journal of Physical Chemistry* **72**, 2742–2749.

Fliermans, C.B. & Hazen, T.C. (1980) Immunofluorescence of *Aeromonas hydrophila* as measured by fluorescence photometric microscopy. *Canadian Journal of Microbiology* **26**, 161–168.

Foon, K.A., Schroff, R.W. & Gale, R.P. (1982) Surface markers on leukemia and lymphoma cells; recent advances. *Blood* **60**, 1–19.

Forman, D.S. & Turriff, D.E. (1981) Video intensification microscopy (VIM) as an aid to routine fluorescence microscopy. *Histochemistry* **71**, 203–208.

Foskett, J.K. (1988) Simultaneous Nomarski and fluorescence imaging during video microscopy of cells. *American Journal of Physiology* **255**, C566–C571.

Franke, H., Barlow, C.H. & Chance, B. (1980) Surface fluorescence of reduced pyridine nucleotide of the perfused rat kidney: interrelation between metabolic and functional states. *Contributions to Nephrology* **19**, 240–247.

Franklin, A.L. & Filion, W.G. (1985) A new technique for retarding fading of fluorescence: DPX-BME. *Stain Technology* **60**, 125–135.

Franzeck, U.K., Isenring, G., Frey, J. & Bollinger, A. (1983a) Video-densitometric pattern recognition of Na-fluorescein diffusion in nail fold capillary areas of patients with acrocyanosis, primary vasospastic and secondary Raynaud's phenomenon. *International Angiology* **2**, 143–152.

Franzeck, U.K., Isenring, G., Frey, J., Jäger, K., Mahle, F. & Bollinger, A. (1983b) Eine Apparatur zur dynamischen intravitalen Videomikroskopie. *Vasa* **12**, 233–238.

Freeman, H. (1963) *Introduction to statistical inference*. 445 pp. Addison-Wesley, Reading, MA.

Freitas, M.I., Giordano, P.A. & Bottiroli, G. (1981) Improvement in microscope photometry by voltage to frequency conversion: analogue measurement and digital processing. *Journal of Microscopy* **124**, 211–218.

Frey-Wyssling, A. (1964) Ultraviolet and fluorescence optics of lignified cell walls. In: Zimmerman, M.H. (Ed.) *Formation of wood in forest trees*. Second symposium of the M.M. Cabot Foundation, 1963, pp. 153–167. Academic Press, New York.

Fujita, S. (1973) DNA cytofluorometry on large and small cell nuclei stained with pararosaniline Feulgen. *Histochemie* **36**, 193–199.

Fujita, S. & Fukuda, M. (1974) Irradiation of specimens by excitation light before and after staining with pararosaniline Feulgen: a new method to reduce non-specific fluorescence in cytofluorometry. *Histochemistry* **40**, 59–67.

Fukuda, M. (1983) Application of the microcomputer in multiparametric fluorescence cytophotometry. *Journal of Histochemistry and Cytochemistry* **31**, 241–243.

Fukuda, M., Böhm, N. & Fujita, S. (1978) Cytophotometry and its biological application. *Progress in Histochemistry and Cytochemistry* **11**, (1). 119 pp.

Fukuda, M., Hoshino, K., Naito, M. & Morita, T. (1982) A fluorescence cytophotometer operated under computer control for multi-parameter cell analysis. *Histochemistry* **76**, 1–13.

Fukuda, M., Tsuchihashi, Y., Takamatsu, T., Nakanishi, K. & Fujita, S. (1980) Fluorescence fading and stabilization in cytofluorometry. *Histochemistry* **65**, 269–276.

Fulwyler, M.J. (1965) Electronic separation of biological cells by volume. *Science* **150**, 910–911.

Gabler, F., Gubisch, W., Lipp, W. & Rüker, O. (1960) Das Reflexphotometer. Ein neues Konstruktionsprinzip für Mikroskop-Spekralphotometer. *Mikroskopie* **15**, 218–226.

Gahrton, G. & Yataganas, X. (1976) Quantitative cytochemistry of glycogen in blood cells. Methods and clinical application. *Progress in Histochemistry and Cytochemistry* **9**, 1–30.

Gains, N. & Dawson, A.P. (1979) Relevance of the approximately hyperbolic relationship between fluorescence and concentration to the determination of quantum efficiencies. *The Analyst* **104**, 481–490.

Galassi, L. (1990) Fluorescence discrimination in a reflecting environment. *Journal of Microscopy* **157**, 181–186.

Galbraith, W., Geyer, B. & David, G.B. (1975) The Denver universal microspectroradiometer (DUM). II. Computer configuration and modular programming for radiometry. *Journal of Microscopy* **105**, 237–264.

Galjaard, H., Hoogeveen, A., Keijzer, W., de Wit-Verbeek, E. & Vlek-Noot, C. (1974) The use of quantitative cytochemical analyses in rapid prenatal detection and somatic cell genetic studies of metabolic diseases. *Histochemical Journal* **6**, 491–509.

Ganse, H. (1977) Polarisations Mikroskopie. In: Beyer, H. (Ed.) *Handbuch der Mikroskopie*, 2nd edn, pp. 193–223. VEB Verlag Technik, Berlin.

Garcia, A.M. (1962) Studies on DNA in leucocytes and related cells of mammals. II. On the Feulgen reaction and two-wavelength microspectrophotometry. *Histochemie* **3**, 178–194.

Garcia, A.M. & Iorio, R. (1966) Potential sources of error in two-wavelength cytophotometry. In: Wied, G.L. (Ed.) *Introduction to quantitative cytochemistry*, pp. 215–234. Academic Press, New York & London.

Garland, P.B. & Moore, C.H. (1979) A possible probe for measurements of slow rotational activity. *Biochemical Journal* **183**, 561–572

Gawlitta, W., Stockem, W., Wehland, J. & Weber, K. (1980) Pinocytosis and locomotion of amoebae. XV. Visualization of Ca^{++}-dynamics by chlortetracycline (CTC) fluorescence during induced pinocytosis in living *Amoeba proteus*. *Cell and Tissue Research* **213**, 9–20.

Geacintov, N.E., Van Nostrand, F. & Becker, J.F. (1974) Polarized light spectroscopy of photosynthetic membranes in magneto-oriented whole cells and chloroplasts. Fluorescence and dichroism. *Biochemica et biophysica acta* **347**, 443–463.

Geber, G. & Hasibeder, G. (1980) Cytophotometrische Bestimmung von DNA-Mengen: Vergleich einer neuen DAPI-Fluoreszenzmethode mit Feulgen-Absorptionsphotometrie. [Cytophotometric estimation of DNA contents: comparison of a new DAPI fluorescence method with Feulgen absorbance photometry.] *Microscopica acta* Supplementum **4**, 31–35.

Geel, F. van, Smith, B.W., Nicolaissen, B. & Wineforder, J.D. (1984) Epifluorescence microscopy with a pulsed nitrogen tunable dye laser source. *Journal of Microscopy* **133**, 141–148.

Geyer, M.A., Dawsey, W.J. & Mandell, A.J. (1978) Fading: a new cytofluorimetric measure quantifying serotonin in the presence of catecholamines at the cellular level in brain. *Journal of Pharmacology and Experimental Therapeutics* **207**, 650–667.

Ghetti, F., Colombetti, G., Lenci, F., Campani, E., Polacco, E. & Quaglia, M. (1985) Fluorescence of *Euglena gracilis* photoreceptor pigment: an *in vivo* microspectrofluorometric study. *Photochemistry and Photobiology* **42**, 29–33.

Giangaspero, F., Chieco, P., Lisignoli, G. & Burger, P.C. (1987) Comparison of cytologic composition with microfluorometric DNA analysis of the glioblastoma multiforme and anaplastic astrocytoma. *Cancer* **60**, 59–65.

Gijzel, P. van (1966) Die Fluoreszenz-Photometrie von Mikrofossilien mit dem Zweistrahl-Mikroskop-photometer nach Berek. *Leitz Mitteilungen für Wissenschaft und Technik* **3**, 206–214. Also English edition: Fluorescence photometry of microfossils with the Berek two-beam microscope photometer. *Leitz Scientific and Technical Information* **1**, 174–182. (Colour plate.)

Gijzel, P. van (1971) Review of the UV fluorescence photometry of fresh and fossil exines and exosporia. In: Brooks, J., Grant, P.R., Muir, M.D., van Gijzel, P. & Shaw, G. (Eds.) *Sporopollenin*, pp. 659–685. Academic Press, London.

Gijzel, P. van (1973) *Topics in UV-fluorescence microspectrophotometry II. Fresh and fossil plant substances.* Katholisch Universität, Nijmegen.

Gijzel, P. van (1975) Polychromatic UV-fluorescence microphotometry of fresh and fossil plant substances with special reference to the location and identification of dispersed organic material in rocks. In: Alpern, B. (Ed.) *Pétrographie Organique et Potentiel Pétrolier*, pp. 67–91. CNRS, Paris.

Gijzel, P. van (1977) Die Fluoreszenz-Mikroskopie einiger Pflanzenfossilien. *Courier Forschungsinstitut Senekenberg* (Senekenburgische Naturforschende Gesellschaft), Frankfurt am Main, **24**, 92–100.

Gijzel, P. van (1978) Recent developments in the application of quantitative fluorescence microscopy in palynology and paleobotany. *Annales des Mines de Belgique* No. 7.d, pp. 165–181.

Gill, D. (1979) Inhibition of fading in fluorescence in microscopy of fixed cells. *Experientia* **35**, 400–401.

Gillis, C.N., Schneider, F.H., Van Orden, L.S. & Giarman, N.J. (1966) Biochemical and microfluoro-metric studies of norepinephrine redistribution accompanying sympathetic nerve stimulation. *Journal of Pharmacology and Experimental Therapeutics* **151**, 46–54.

Giloh, H. & Sedat, J.W. (1982) Fluorescence microscopy; reduced photobleaching of rhodamine and fluorescein protein conjugates by *n*-propyl gallate. *Science* **217**, 1252–1255.

Giordano, P., Bottiroli, G., Prenna, G., Doglia, S. & Baldini, G. (1977) Low temperature microspectro-fluorometry: design of a 'cold chamber'. *Journal of Microscopy* **112**, 95–101.

Giordano, P.A., Prosperi, E. & Bottiroli, G. (1984) Primary fluorescence of rat muscle after CO_2 laser thermal injury. *Lasers in Surgery and Medicine* **4**, 271–278.

Goff, L.J. & Coleman, A.W. (1984) Elucidation of fertilization and development in a red alga by quantitative DNA microspectrofluorometry. *Developmental Biology* **102**, 173–194.

Golden, J.F. & West, S.S. (1974) Fluorescence spectroscopic and fading behavior of Ehrlich's hyper-diploid mouse ascites tumor cells supravitally stained with acridine orange. *Journal of Histochemistry and Cytochemistry* **22**, 495–505.

Golden, J.F., West, S.S., Shingleton, H.M., Murad, T.M. & Echols, C.K. (1979) A screening system for cervical cancer cytology. *Journal of Histochemistry and Cytochemistry* **27**, 522–528.

Goldman, M. (1960) Antigenic analysis of *Entamoeba histolytica* by means of fluorescent antibody. I. Instrumentation for microfluorimetry of stained amebae. *Experimental Parasitology* **9**, 25–36.

Goldman, M. (1967) An improved microfluorimeter for measuring brightness of fluorescent antibody reactions. *Journal of Histochemistry and Cytochemistry* **15**, 38–45.

Goldman, M. & Carver, R.K. (1961) Microfluorometry of cells stained with fluorescent antibody. *Experimental Cell Research* **23**, 265–280.

Goldman, R.N. & Weinberg, J.S. (1985) *Statistics, an introduction.* 773 pp. Prentice-Hall, Englewood Cliffs, NJ.

Gordon, L.K. & Parker, C.W. (1981) A microfluorometric assay for the measurement of *de novo* DNA synthesis in individual cells. *Journal of Immunology* **127**, 1634–1639.

Gorman, M. & Winkler, P.W. (Eds.) (1978) *Anglo-American cataloguing rules*, 2nd edn. 619 pp. The Library Association, London.

Gratama, J.W., Naipal, A.M.I.H., Oljans, P.J., Zwaan, F.E., Verdonck, L.F., de Witte T., Vossen, J.M.J.J., Bolhuis, R.H.L., de Gast, G.C. & Jansen, J. (1984) Early shifts in the ratio between T4[+] and

T8$^+$ T lymphocytes correlates with the occurrence of acute graft-versus-host disease. *Blood* **63**, 1416–1423.

Green, W.B. (1983) *Digital image processing, a systems approach.* Van Nostrand-Reinhold, New York.

Grynkiewicz, G., Poenie, M. & Tsien, R.Y. (1985) A new generation of Ca^2+ indicators with greatly improved fluorescence properties. *Journal of Biological Chemistry* **260**, 3440–3450.

Grzywacz, J. (1967) Difference between prompt and delayed fluorescence spectra. *Nature* **213**, 385–386.

Guilbault, G.G. (1973) (Ed.) *Practical fluorescence: theory, methods and techniques.* Marcel Dekker, New York.

Gustavsson, B. (1980) Cytofluorometric analysis of anaphylactic secretion of 5-hydroxytryptamine and heparin from rat mast cells. *International Archives of Allergy and Applied Immunology* **63**, 121–128.

Gustavsson, B. & Enerbäck, L. (1978) Cytofluorometric quantitation of 5-hydroxytryptamine and heparin in individual mast cell granules. *Journal of Histochemistry and Cytochemistry* **26**, 47–54.

Gustavsson, B. & Enerbäck, L. (1980) A cytofluorometric analysis of polymer-induced mast cell secretion. *Experimental Cell Biology* **48**, 15–30.

Gutjahr, C.C.M. (1983) Introduction to incident-light microscopy of oil and gas source rocks. *Geologie en Mijnbouw* **62**, 417–425.

Haaijman, J.J. & Dalen, J.P.R. van (1974). Quantification in immunofluorescence microscopy. A new standard for fluorescein and rhodamine emission measurement. *Journal of Immunological Methods* **5**, 359–374.

Haaijman, J.J. & Slingerland-Teunissen, J. (1978) Equipment and preparative procedures in immunofluorescence microscopy: quantitative studies. In: Knapp, W., Holubar, K. & Wick, G. (Eds.) *Sixth international conference on immunofluorescence and related staining techniques, Vienna,* pp. 11–29. Elsevier/North Holland, New York & Amsterdam.

Haaijman, J.J. & Wijnants, F.A.C. (1975) Inexpensive automation of the Leitz Orthoplan microfluorometer using pneumatic components. *Journal of Immunological Methods* **7**, 255–270.

Haitinger, M. (1938) *Fluoreszenzmikroskopie. Ihre Anwendung in der Histologie und Chemie.* Akademische Verlagsgesellschaft M.B.H., Leipzig.

Haitinger, M. (1959) *Fluoreszenz-Mikroskopie,* 2nd edn, ed. Eisenbrand, J. & Werth, G. Akademische Verlagsgesellschaft Geest & Portig K.-G, Leipsig.

Håkanson, R. & Sundler, F. (1971) Formaldehyde condensation. A method for the fluorescence microscopic demonstration of peptides with NH_2-terminal tryptophan residues. *Journal of Histochemistry and Cytochemistry* **19**, 477–482.

Hald, A. (1952) *Statistical theory with engineering applications.* 783 pp. J. Wiley & Sons, New York.

Hamada, S. & Fujita, S. (1983) DAPI staining improved for quantitative cytofluorometry. *Histochemistry* **79**, 219–226.

Hamberger, B., Ritzén, M. & Wersall, J. (1966) Demonstration of catecholamines and 5-hydroxytryptamine in the human carotid body. *Journal of Pharmacology and Experimental Therapeutics* **152**, 197–201.

Hamilton, D.K. & Wilson, T. (1986) Scanning optical microscopy by objective lens scanning. *Journal of Physics E: Scientific Instruments* **19**, 52–54.

Hamilton, D.K., Wilson, T. & Sheppard, C.J.R. (1981) Experimental observations of the depth-discrimination properties of scanning microscopes. *Optics Letters* **6**, 625–626.

Hamperl, H. (1934) Die Fluoreszenzmikroskopie menschlicher Gewebe. *Virchows Archiv für pathologische Anatomie* **292**, 1–51.

Hargittai, P.T., Ginty, D.D. & Lieberman, E.M. (1987) A pyrene fluorescence technique and microchamber for measurement of oxygen consumption of single isolated axons. *Analytical Biochemistry* **163**, 418–426.

Haugland, R.P. (1989) *Handbook of fluorescent probes and research chemicals.* Molecular Probes, Eugene, OR.

Hazum, E., Chang, K.J. & Cuatrecasas, P. (1979) Opiate (encephalin) receptors of neuroblastoma cells: occurrence in clusters on the cell surface. *Science* **206**, 1077–1079.

Hedley, D.W., Friedlander, M.L., Taylor, I.W., Rugg, C.A. & Musgrove, E.A. (1983) Method for analysis of cellular DNA content of paraffin embedded pathologic material using flow cytometry. *Journal of Histochemistry and Cytochemistry* **31**, 1333–1335.

Heimer, G.V. & Taylor, C.E.D. (1974) Improved mountant for immunofluorescence preparations. *Journal of Clinical Pathology* **27**, 254–256.

Heiple, J.M. & Taylor, D.L. (1982) pH changes in pinosomes and phagosomes in the ameba, *Chaos carolinensis*. *Journal of Cell Biology* **94**, 143–149.

Hemstreet, G.P., West, S.W., Weems, W.L., Echols, C.K., McFarland, S., Lewin, J. & Lindseth, G. (1983) Quantitative fluorescence measurements of AO-stained normal and malignant bladder cells. *International Journal of Cancer* **31**, 577–585.

Hengartner, H. (1961) Die Fluoreszenzpolarisation der verholzten Zellwand. *Holz als Roh- und Werkstoff* **19**, 303–309.

Henschen, A. & Olson, L. (1983) Hexachlorophene-induced degeneration of adrenergic nerves: application of quantitative image analysis to Falck-Hillarp fluorescence histochemistry. *Acta neuropathologica* **59**, 109–114.

Herman, B. & Albertini, D. (1984) A time-lapse video image intensification analysis of cytoplasmic organelle movements during endosome translocation. *Journal of Cell Biology* **98**, 565–576.

Herrman, J. & Wilhelmi, B. (1987) *Laser für ultrakurze Lichtimpulse*. Akademie-Verlag, Berlin. (English translation by J. Grossman & W. Rudolph, *Lasers for ultrashort light pulses*. 302 pp. North-Holland, Amsterdam.)

Heslop-Harrison, J. & Heslop-Harrison, Y. (1970) Evaluation of pollen viability by enzymatically induced fluorescence: intracellular hydrolysis of fluorescein diacetate. *Stain Technology* **45**, 115–120.

Hirschberg, J.G., Wouters, A.W., Kohen, E., Kohen, C., Thorell, B., Eisenberg, B., Salmon, J.M. & Ploem, J.S. (1979) A high resolution grating microspectrofluorometer with topographic option for studies in living cells. In: Talmi, Y. (Ed.) *Multichannel image detectors*, ACS Symposium Series, No. **102**, pp. 263–289. American Chemical Society.

Hirschfeld, T. (1979) Fluorescence background discrimination by prebleaching. *Journal of Histochemistry and Cytochemistry* **27**, 96–101.

Hobson, G.S. (1978) *Charge transfer devices*. 207 pp. Edward Arnold, London.

Hollis, A.J. (1987) Photoelectric photometry: pulse counting or direct current? *Journal of the British Astronomical Association* **97**, 120–122.

Homann, W. (1972) Zum spektralen Fluoreszenz-Verhalten des Sporinits in Kohlen-Anschliffen und seine Bedeutung für die Inkohlungsgrad-Bestimmung. *Report to the 25th meeting of the International Commission for Coal Petrology*, Belgrade.

Hook, G.R. & Odeyale, C.O. (1989) Confocal scanning fluorescence microscopy: a new method for phagocytosis research. *Journal of Leukocyte Biology* **45**, 277–282.

Horikawa, Y., Yamamoto, M. & Dosaka, S. (1987) Laser scanning microscope: differential phase images. *Journal of Microscopy* **148**, 1–10.

Horster, M.F., Wilson, P.D. & Gundlach, H. (1983) Direct evaluation of fluorescence in single renal epithelial cells using a mitochondrial probe (DASPMI). *Journal of Microscopy* **132**, 143–148.

Hovnanian, H.P., Brennan, T.A. & Botan, E.A. (1964) Quantitative rapid immunofluorescence microscopy. I. Instrumentation. *Journal of Bacteriology* **87**, 473–476.

Huff, J.C., Weston, W.L. & Wanda, K.D. (1982) Enhancement of specific immunofluorescent findings with use of a para-phenylenediamine mounting buffer. *Journal of Investigative Dermatology* **78**, 449–450.

Huitfeldt, H.S., Spangler, E.F., Baron, J. & Poirier, M.C. (1987) Microfluorometric determination of DNA addducts in immunofluorescent-stained liver tissue from rats fed 2-acetylaminofluorene. *Cancer Research* **47**, 2098–2102.

Hutz, R.J, DeMayo, F.J. & Dukelow, W.R. (1985) The use of vital dyes to assess embryonic viability in the hamster, *Mesocricetus auratus*. *Stain Technology* **60**, 163–168.

Inoki, S., Osaki, H. & Furuya, M. (1979) *In situ* microspectrofluorometry of nuclear and kinetoplast DNA in *Trypanosoma gambiense*. *Zentralblatt für Bakteriologie und Hygiene, I. Abt. Orig. A* **244**, 327–330.

Inoué, S. (1981) Video image processing greatly enhances contrast, quality, and speed in polarization-based microscopy. *Journal of Cell Biology* **89**, 346–356.

Inoué, S. (1986) *Video microscopy*. 584 pp. Plenum Press, New York & London.

Inoué, S. (1988) Progress in video microscopy. *Cell Motility and the Cytoskeleton* **10**, 13–17.

Isings, J. (1966) Combined fluorescence dichroism and polarising microscopy of cotton fibres under stress. *The Microscope* **15**, 71–79.

Jacob, H. (1974) Fluoreszenz-Mikroskopie und -Photometrie der organischen Substanz von Sedimenten und Boden. In: Freund, H. (Ed.) *Handbuch der Microskopie in der Technik*, vol. **IV**, part 2, pp. 369–391. Umschau Verlag, Frankfurt (Main).

Jacobson, K., Elson, E., Koppel, D. & Webb, W. (1983) International workshop on the application of fluorescence photobleaching techniques to problems in cell biology. *Federation Proceedings* **42**, 72–79.

James, J. (1976) *Light microscopic techniques in biology and medicine*. 336 pp. Martinus Nijhoff, Amsterdam.

Jarvis, L.R. (1988) Microcomputer video image analysis. *Journal of Microscopy* **150**, 83–97.

Ji, S., Chance, B., Nishiki, K., Smith, T. & Rich, T. (1979) Micro-light guides: a new method for measuring tissue fluorescence and reflectance. *American Journal of Physiology* **236**, C144–C156.

Ji, S., Lemasters, J.J., Christenson, V. & Thurman, R.G. (1982) Periportal and pericentral pyridine nucleotide fluorescence from the surface of the perfused liver: evaluation of the hypothesis that chronic treatment with ethanol produces pericentral hypoxia. *Proceedings of the National Academy of Sciences, USA* **79**, 5415–5419.

Johannisson, E. & Thorell, B. (1977) Mithramycin fluorescence for quantitative determination of deoxyribonucleic acid in single cells. *Journal of Histochemistry and Cytochemistry* **25**, 122–128.

Johnson, G.D., Davidson, R.S., McNamee, K.C., Russell, G., Goodwin, D. & Holborow, E.J. (1982) Fading of immunofluorescence during microscopy: a study of the phenomenon and its remedy. *Journal of Immunological Methods* **55**, 231–242.

Johnson, G.D. & Nogueira Araujo, G.M. de C. (1981) A simple method of reducing the fading of immunofluorescence during microscopy. *Journal of Immunological Methods* **43**, 349–350.

Jongkind, J.F., Ploem, J.S., Revser, A.J.J. & Galjaard, H. (1974) Enzyme assays at the single cell level using a new type of microfluorimeter. *Histochemistry* **40**, 221–229.

Jongkind, J.F., Verfkerk, A., Visser, J.W. & Dongen, J.M. van (1982) Isolation of autofluorescent "aged" human fibroblasts by flow sorting. *Experimental Cell Research* **138**, 409–417.

Jongsma, A.P.M., Hijmans, W. & Ploem, J.S. (1971) Quantitative immunofluorescence. Standardization and calibration in microfluorometry. *Histochemie* **25**, 329–343.

Jonsson, G. (1967) Fluorescence method for histochemical demonstration of monoamines. VII. Fluorescence studies on biogenic amines and related compounds condensed with formaldehyde. *Histochemie* **8**, 288–296.

Jonsson, G. (1971) Quantitation of fluorescence of biogenic amines. *Progress in Histochemistry and Cytochemistry* **2**, (4). 36 pp.

Jonsson, G., Einarsson, P., Fuxe, K. & Hallman, H. (1975) Microspectrofluorometric analysis of the formaldehyde induced fluorescence in midbrain raphe neurons. *Medical Biology* **53**, 25–39.

Jonsson, G. & Ritzén, M. (1966) Microspectrofluorimetric identification of metaraminol in sympathetic adrenergic neurons. *Acta physiologica scandinavica* **67**, 505–513.

Jonsson, G. & Sachs, C. (1971) Microspectrofluorimetric identification of *m*-hydroxyphenyl-ethylamines (*m*-tyramines) in central and peripheral monoamine neurons. *Histochemie* **25**, 208–216.

Jonsson, G. & Sandler, M. (1969) Fluorescence of indolylethylamines condensed with formaldehyde. *Histochemie* **17**, 207–212.

Jotz, M.M., Gill, J.E. & Davis, D.T. (1976) A new optical multichannel microspectrofluorometer. *Journal of Histochemistry and Cytochemistry* **24**, 91–96.

Jovin, T. (1979) Fluorescence polarization and energy transfer: theory and application. In: Melamed, M.R., Mullaney, P.F. & Mendelsohn, M.L. (Eds.) *Flow cytometry and sorting*, pp. 137–165. John Wiley, New York.

Kachel, V. (1979) Electrical resistance pulse sizing (Coulter sizing). In: Melamed, M.R., Mullaney, P.F. & Mendelsohn, M.L. (Eds.) *Flow cytometry and sorting*, pp. 61–104. John Wiley, New York.

Kamentski, L.A., Melamed, M.R. & Derman, H. (1965) Spectrofotometer: new instrument for ultrarapid cell analysis. *Science* **150**, 630–631.

Kapitza, H.G., McGregor, G. & Jacobson, K.A. (1985) Direct measurement of lateral transport in membranes by using time-resolved spatial photometry. *Proceedings of the National Academy of Sciences, USA* **82**, 4122–4126.

Kask, P., Piksarv, P. & Mets, Ü. (1985) Fluorescence correlation spectroscopy in the nanosecond time range: photon antibunching in dye fluorescence. *European Biophysics Journal* **12**, 163–166.

Kassotis, J., Steinberg, S.F., Ross, S., Bilezikian, J.P. & Robinson, R.B. (1987) An inexpensive dual-excitation apparatus for fluorescence microscopy. *Pflügers Archiv* **409**, 47–51.

Kasten, F.H. (1958) Additional Schiff-type regents for use in cytochemistry. *Stain Technology* **33**, 39–45.

Kasten, F.H. (1983) The development of fluorescence microscopy up through World War II. In: Clark, G. & Kasten, F.H. (Eds.) *History of staining*, 3rd edn, Chapter 23, pp. 147–185. Williams & Wilkins, Baltimore & London.

Kasten, F.H. (1989) The origins of modern fluorescence microscopy and fluorescent probes. In: Kohen, E. & Hirschberg, J.G. (Eds.) *Cell structure and function by microspectrofluorometry*, pp. 3–50. Academic Press, San Diego.

Kaufman, G.I., Nester, J.F. & Wasserman, D.E. (1971) An experimental study of lasers as excitation sources for automated fluorescent antibody instrumentation. *Journal of Histochemistry and Cytochemistry* **19**, 469–476.

Kearns, D.R. (1971) Physical and chemical properties of singlet molecular oxygen. *Chemical Reviews* **71**, 393–427.

Kemp, C.L., Doyle, G. & Anderson, R. (1979) Microfluorometric measurement of DNA in *Eudorina elegans* and *E. californica* (Chlorophyceae). *Journal of Phycology* **15**, 464–465.

Kemplay, J.R. (1962) The electronic requirements of a spectrofluorometer. *Electronic Engineering* **34**, 820–823.

Kerker, M. (1969) *The scattering of light and other electromagnetic radiation*. Academic Press, New York.

Kielland, J. (1941) Method and apparatus for counting blood corpuscles. US Patent No. 2 369 577. Filed 12 May 1941, issued 13 February 1945.

Kirsch, B., Voigtman, E. & Wineforder, J.D. (1985) High-sensitivity laser fluorometer. *Analytical Chemistry* **57**, 2009–2011.

Klig, V., Demirjian, C. & Pungaliya, P. (1976) Microspectrofluorometry in the study of biogenic amines: automatic correction of excitation spectra. *Journal of Microscopy* **107**, 173–176.

Knapp, W. & Ploem, J.S. (1974). Microfluorometry of antigen–antibody interactions in immunofluorescence using the defined antigen substrate spheres (DASS) system. Sensitivity, specificity and variables of the method. *Journal of Immunological Methods* **5**, 259–273.

Knoester, A. & Brakenhoff, G.J. (1990) Applications of confocal microscopy in industrial solid materials: some examples and a first evaluation. *Journal of Microscopy* **157**, 105–114.

Kohen, E. (1963) A flow chamber for the differential microfluorimeter of Chance and Legallais. Preliminary work with glass-grown ascites cells. *Biochimica et biophysica acta* **75**, 139–142.

Kohen, E. (1964a) Oxidation–reduction kinetics of pyridine nucleotide. A microfluorimetric study in insect spermatids and ascites cells. *Experimental Cell Research* **35**, 26–37.

Kohen, E. (1964b) Pyridine nucleotide compartmentalization in glass-grown ascites cells. *Experimental Cell Research* **35**, 303–316.

Kohen, E. & Hirschberg, J.G. (Eds.) (1989) *Cell structure and function by microspectrofluorometry*. 465 pp. Academic Press, San Diego.

Kohen, E., Hirschberg, J.G., Fried, M., Kohen, C., Santus, R., Reyftmann, J.P., Morliere, P., Schactschable, D.O., Mangel, W.F., Shapiro, B.G. & Prince, J. (1987) Microspectrofluorometric study of cell structure and function. In: Etz, E.S., Milanovitch, F.P., Dhamelincourt, P. & Cook, B.W. (Eds.) 22nd annual meeting of the Microbeam Analysis Society, Kailua-Kona, Hawaii, July 1987.

Kohen, E., Hirschberg, J.G. & Prince, J. (1989) Spatiotemporal mapping of fluorescence parameters in cells treated with toxic chemicals: the cell's detoxification apparatus. *Optical Engineering* **28**, 222–231.

Kohen, E., Hirschberg, J.G. & Rabinovitch, A. (1985) Applications of microspectrofluorometry to metabolic control, cell physiology and pathology. In: Cowden, R.R. & Harrison, F.W. (Eds.) *Advances in microscopy*, pp. 45–72. Alan R. Liss, New York.

Kohen, E., Hirschberg, J.G., Thorell, B., Kohen, C. & Mansell, J. (1982) Applications of microspectrof-luorometry to the study of the living cell. In: Heinrich, K.R.G. (Ed.) *Microbeam Analysis – 1982* (17th annual meeting of the Microbeam Analysis Society, Washington, DC, August 1982), pp. 321–329. San Francisco Press, Berkeley, CA.

Kohen, E. & Kohen, C. (1966) A study of mitochondrial–extramitochondrial interactions in giant tissue culture cells by microfluorimetry-microelectrophoresis. *Histochemie* 7, 339–347.

Kohen, E. & Kohen, C. (1977) Rapid automated multichannel microspectrofluorometry. *Experimental Cell Research* 107, 261–268.

Kohen, E., Kohen, C. & Hirschberg, J.G. (1983) Microspectrofluorometry of carcinogens in living cells. *Histochemistry* 79, 31–52.

Kohen, E., Kohen, C., Hirschberg, J.G., Wouters, A., Bartick, P.R., Thorell, B., Bereiter-Hahn, J., Meda, P., Rabinovitch, A., Mintz, D. & Ploem, J.S. (1981a) Examination of single cells by microspectrophoto-metry and microspectrofluorimetry. *Techniques in Cellular Physiology* P103, 1–28.

Kohen, E., Kohen, C., Hirschberg, J.G., Wouters, A., Thorell, B., Westerhoff, H.V. & Charylulu, K.K.N. (1983) Metabolic control and compartmentation in single living cells. *Cell Biochemistry and Function* 1, 3–16.

Kohen, E., Kohen, C. & Jenkins, W. (1966) The influence of microelectrophoretically introduced metabolites on pyridine nucleotide reduction in giant culture ascites cells. *Experimental Cell Research* 44, 175–194.

Kohen, E., Kohen, C., Nordberg, L., Thorell, B. & Akerman, A. (1967) A beam splitter supplemented Chance–Legallais microfluorometer for kinetic studies synchronous with cell manipulations. *Digest of the 7th international conference on medical and biological engineering*, p. 274. Stockholm.

Kohen, E., Kohen, C. & Thorell, B. (1968a) A comparative study of pyridine nucleotide metabolism in yeast and mammalian cells by microfluorimetry-microelectrophoresis. *Histochemie* 12, 95–106.

Kohen, E., Kohen, C. & Thorell, B. (1968b) Kinetics of NAD reduction in the nucleus and cytoplasm. *Histochemie* 16, 170–185.

Kohen, E., Kohen, C. & Thorell, B. (1969) Use of microfluorimetry to study the metabolism of intact cells. *Biomedical Engineering* 4, 544–559.

Kohen, E., Kohen, C., Thorell, B. & Wagner, G. (1973) Quantitative aspects of rapid microfluorometry for the study of enzyme reactions and transport mechanisms in single living cells. In: Thaer, A. & Sernetz, M. (Eds.) *Fluorescence techniques in cell biology*, pp. 207–218. Springer-Verlag, Berlin, Heidelberg & New York.

Kohen, E. & Legallais, V. (1965) Microelectrophoresis and microinjection assembly for the Chance–Legallais microfluorimeter. *Review of Scientific Instruments* 36, 1890–1891.

Kohen, E., Legallais, V. & Kohen, C. (1966) An introduction to microelectrophoresis and microinjection techniques in microfluorimetry. *Experimental Cell Research* 41, 223–226.

Kohen, E., Thorell, B., Hirschberg, J.G., Wouters, A.W., Kohen, C., Bartick, P.R., Salmon, J.-M., Viallet, P., Schachtschabel, D.O., Rabinovitch, A., Mintz, D., Meda, P., Westerhoff, H., Nestor, J. & Ploem, J.S. (1981b) Microspectrofluorometric procedures and their applications in biological systems. In: Wehry, E.L. (Ed.) *Modern fluorescence spectroscopy*, vol. III, Chapter 7, pp. 295–346. Plenum Press, New York.

Kolb, M.J. & Bourne, W.M. (1986) Supravital fluorescent staining of the corneal endothelium with acridine orange and ethidium bromide. *Current Eye Research* 7, 485–494.

Kopac, M.J. (1964) Micromanipulators. Principles of design, operation and application. In: Nastuk, W.L. (Ed.) *Physical techniques in biological research*, vol. 5 (Electrophysiological methods) Part A, pp. 191–233. Academic Press, New York.

Koppel, D.E. (1979) Fluorescence redistribution after photobleaching: a new multi-point analysis of membrane translational dynamics. *Biophysical Journal* 28, 281–291.

Koppel, D.E., Carlson, C. & Smilowitz, H. (1989) Analysis of heterogeneous fluorescence photobleaching by video kinetics imaging: the method of cumulants. *Journal of Microscopy* 155, 199–206.

Koppel, D.E., Primakoff, P. & Myles, D.G. (1986) Fluorescence photobleaching analysis of cell surface regionalization. In: Taylor, D.L, Waggoner, A.S., Murphy, R.F., Lanni, F. & Birge, R.R. (Eds.)

Applications of fluorescence in the biomedical sciences, pp. 477–497. Alan R. Liss, New York.

Koretsky, A.P., Katz, L.A. & Balaban, R.S. (1987) Determination of pyridine nucleotide fluorescence from the perfused heart using an internal standard. *American Journal of Physiology* **253**, H856–H862.

Kortüm, G. (1962) *Kolorimetrie, Photometrie und Spektrometrie*. Springer-Verlag, Berlin, Heidelberg & New York. (Cited by Rigler, R. (1966) Microfluorometric characterisation of intracellular nucleic acids and nucleoproteins by Acridine Orange. *Acta physiologica scandinavica* **67**, Supplementum **267**.)

Krasnovsky, A.A. (1982) Delayed fluorescence and phosphorescence of plant pigments. *Photochemistry and Photobiology* **36**, 733–741.

Kreyszig, E. (1970) *Introductory mathematical statistics*. 470 pp. Wiley, New York.

Kucera, P. & Deribaupierre, Y. (1980) Evaluation of the NAD redox states in a living chick embryo using a computer-controlled double-beam scanning microfluorometry. *Microscopica acta* **S4**, 283–287.

Kunz, C. (1964) The use of the immunofluorescent method and microphotometry for the differentiation of arboviruses. *Virology* **24**, 672–674.

Kurtz, I. & Balaban, R.S. (1985) Fluorescence emission spectroscopy of 14-dihydroxyphthalonitrile. A method for determining intracellular pH in cultured cells. *Biophysical Journal* **48**, 499–508.

Lahmy, S., Salmon, J.M. & Viallet, P. (1984) Quantitative microspectrofluorimetry study of the blocking effect of 6-aminochrysene on benzo(*a*)pyrene metabolism using single living cells. *Toxicology* **29**, 345–356.

Lampton, M. (1981) The microchannel image intensifier. *Scientific American* **245**, 46–55.

Lanier, L.L. & Philips, J.H. (1986) Evidence for three types of human cytotoxic lymphocytes. *Immunology Today* **7**, 132–134.

Larsson, P.-A., Goldstein, M. & Dahlström, A. (1984) A new methodological approach for studying axonal transport: cytofluorometric scanning of nerves. *Journal of Histochemistry and Cytochemistry* **32**, 7–16.

Latt, S. (1979) Fluorescent probes of DNA microstructure and synthesis. In: Melamed, M.R., Mullaney, P.F. & Mendelsohn, M.L. (Eds.) *Flow cytometry and sorting*, Chapter 15, pp. 263–284. John Wiley, New York.

Latt, S.A. (1973) Microfluorometric detection of deoxyribonucleic acid (DNA) replication in human metaphase chromosomes. *Proceedings of the National Academy of Sciences, USA* **70**, 3395–3399.

Lautier, D., Anthelme, B., Lahmy, S., Salmon, J.-M. & Viallet, P. (1986) Mise en évidence de l'accumulation intracellulaire de métabolites intermédiaires fluorescents de quelques hydrocarbures aromatiques polycycliques par résolution de spectres de fluorescence de cellules vivantes isolées. *Comptes Rendus de l'Académie des Sciences de Paris, Série III* **8**, 297–302.

Lawrence, M.E. & Possingham, J.V. (1986a) Microspectrofluorometric measurement of chloroplast DNA in dividing and expanding leaf cells of *Spinacia oleracea*. *Plant Physiology* **81**, 708–710.

Lawrence, M.E. & Possingham, J.V. (1986b) Direct measurement of femtogram amounts of DNA in cells and chloroplasts by quantitative microspectrofluorometry. *Journal of Histochemistry and Cytochemistry* **34**, 761–768.

Lee, J.C. & Bahr, G.F. (1983) Microfluorometric studies on chromosomes. Quantitative determination of protein content of Chinese hamster chromosome 1 in situ with and without trypsin digestion. *Chromosoma* **88**, 374–376.

Lee, L., Chen, C.H. & Chiu, L.A. (1986) Thiazole orange: a new dye for reticulocyte analysis. *Cytometry* **7**, 508–517.

Leemann, U. & Ruch, F. (1982) Cytofluorometric determination of DNA base content in plant nuclei and chromosomes by the fluorochromes DAPI and Chromomycin A3. *Experimental Cell Research* **140**, 275–282.

Leskovar, B., Lo, C.C., Hartig, P.R. & Sauer, K. (1976) Photon counting system for subnanosecond fluorescence lifetime measurements. *Review of Scientific Instruments* **47**, 1113–1121.

Letokhov, V.S. (Ed.) (1986) *Laser analytical spectrochemistry*. 430 pp. Adam Hilger, London.

Levi, M., Tarquini, F., Sgorbati, S. & Sparvoli, S. (1986) Determination of DNA content by static cytofluorometry in nuclei released from fixed plant tissue. *Protoplasma* **132**, 64–68.

Lichtensteiger, W. (1970) Katecholaminhaltige Neurone in der neuroendokrinen Steuerung. *Progress in Histochemistry and Cytochemistry* **1**, (4). 92 pp.

Lindvall, O. & Björklund, A. (1974) The glyoxylic acid fluorescence histochemical method: a detailed account of the methodology for the visualization of central catecholamine neurons. *Histochemistry* **39**, 97–127.

Lippert, E., Nägele, W., Seibold-Blankenstein, I., Staiger, U. & Voss, W. (1959) Messung von Fluorescenzspektren [*sic*] mit Hilfe von Spektralphotometern und Vergleichsstandards. *Zeitschrift für analytische Chemie* **170**, 1–18.

Lo, H.B. & Ting, F.T.C. (1972) Fluorescence microscopy for coal petrology. *Geological Society of America Abstracts with Programs* **5**, 715.

Lodja, Z., Ploeg, M. van der & Duijn, P. van (1967) Phosphates of the Naphthol AS series in the quantitative determination of alkaline and acid phosphatase activities "in situ" studied in polyacrylamide membrane model systems and by cytospectrophotometry. *Histochemie* **11**, 13–32.

Loeser, C.N. & Berkley, C. (1954) Electronic quantitation of light absorption and nuclear fluorescence in living cells. *Science* **119**, 410–411.

Löfström, A., Jonsson, G. & Fuxe, K. (1976) Microfluorometric quantification of catecholamine fluorescence in rat median eminence. I. Aspects on the distribution of dopamine and noradrenaline nerve terminals. *Journal of Histochemistry and Cytochemistry* **24**, 415–429.

Löfström, A., Jonsson, G., Wiesel, F.A. & Fuxe, K. (1976) Microfluorimetric quantification of catecholamine fluorescence in rat median eminence. II. Turnover changes in hormonal states. *Journal of Histochemistry and Cytochemistry* **24**, 430–442.

Lowry, O.H. & Passoneau, J.V. (1972) *A flexible system of enzymatic analysis.* Academic Press, New York & London.

Lumsden, J. (1969) *Elementary statistical method.* 166 pp. University of Western Australia Press, Perth, Australia.

Ma, J., Chapman, G.V., Chen, S., Penny, R. & Breit, S.N. (1987) Flow cytometry with crystal violet to detect intracytoplasmic fluorescence in viable human lymphocytes. *Journal of Immunological Methods* **104**, 195–200.

MacInnes, J.W. & Uretz, R.B. (1966) Organization of DNA in dipteran polytene chromosomes as indicated by polarized fluorescence microscopy. *Science* **151**, 689–691.

Maeda, M. (1984) *Laser dyes: properties of organic compounds for dye lasers.* Academic Press, Tokyo, New York & London.

Majidic, O., Bettelheim, P., Stockinger, H., Aberer, W., Liszka, K., Lutz, D. & Knapp, W. (1984) M2, a novel myelomonocytic cell surface antigen and its distribution on leukemic cells. *International Journal of Cancer* **33**, 616–622.

Malin-Berdel, J. & Valet, G. (1980) Flow cytometric determination of esterase and phosphatase activities and kinetics in hematopoietic cells with fluorogenic substrates. *Cytometry* **1**, 222–228.

Mansberg, H.P. & Kusnetz, J. (1966) Quantitative fluorescence microscopy: fluorescent antibody automatic scanning techniques. *Journal of Histochemistry and Cytochemistry* **14**, 260–273.

Marques, D. & Bastos, A.L. (1969) Microfluorometry of tumour cells stained by non-fluorescent dyes and exposed to ultraviolet light. *Proceedings of NATO International Advanced Study Institute Conference on Biological Effects of Visible Light* pp. 63–64. University of Sassari, Sardinia.

Marques, D., Bastos, A.L., Baptista, A.M., Vigário, J.D., Nunes, J.M., Terrinha, A.M. & Silva, J.A.F. (1968) Microfluorometric studies of ultraviolet induced fluorescence in sarcoma 37 tumour cells (ascitic form) of mice supravitally stained by thiazin dyes. *Proceedings of the third international congress of histochemistry and cytochemistry (New York)*, pp. 171–172. Springer-Verlag, Berlin, Heidelberg & New York.

Marques, D. & Rost, F.W.D. (1973) Microspectrofluorometry of ultraviolet inducible fluorescence in supravitally-stained ascites tumour cells. *Histochemical Journal* **5**, 151–156.

Maunder, C. & Rost, F.W.D. (1972) A microfluorometric study of masked metachromasia in the endocrine polypeptide (APUD) series. *Histochemical Journal* **4**, 145–153.

Maxfield, F.R., Schlessinger, J., Schechter, Y., Pastan, I. & Willingham, M.C. (1978) Collection of insulin, EGF, and α_2-macroglobulin in the same patches on the surface of cultured fibroblasts and common internalization. *Cell* **14**, 805–810.

Mayer, R.J. & Novacek, V.M. (1974) A direct recording corrected microspectrofluorometer. *Journal of*

Microscopy **102**, 165–177.

Mayer, R.T. & Thurston, E.L. (1974) An improved method for standardization of microspectrofluor-ometers. *Stain Technology* **49**, 61–64.

Mazzini, G., Bottiroli, G. & Prenna, G. (1975) An electronic device for the automatic correction of fluorescence emission spectra. *Histochemical Journal* **7**, 291–297.

Meech, R. (1981) Microinjection. *Techniques in Cellular Physiology* **P109**, 1–16.

Mehler, L. & Pick, J. (1932) Über ein Mikroskop zur Untersuchung lebenden Gewebes. Vorläufig Mitteillung. *Anatomische Anzeiger* **75**, 234–240.

Melamed, M.R., Lindmo, M.I. & Mendelsohn, M.L. (Eds.) (1989) *Flow cytometry and sorting*, 2nd edn. John Wiley, New York.

Meldolesi, J., Volpe, P. & Pozzan, T. (1988) The intracellular distribution of calcium. *Trends in Neurosciences* **11**, 449–452.

Melhuish, W.H. (1962) Calibration of spectrofluorimeters for measuring corrected emission spectra. *Journal of the Optical Society of America* **52**, 1256–1258.

Mello, M.L.S. & Vidal, B. de C. (1985) Microspectrofluorimetry of the naturally fluorescent substances of the Malpighian tubules of *Triatoma infestans* and *Panstrongylus megistus*. *Acta histochemica et cytochemica* **18**, 365–373.

Mellors, R.C., Glassman, A. & Papanicolaou, G.N. (1952) A microfluorometric scanning method for the detection of cancer cells in smears of exfoliated cells. *Cancer* **5**, 458–468.

Mellors, R.C. & Silver, R. (1951) A microfluorometric scanner for the differential detection of cells: application to exfoliative cytology. *Science* **114**, 356–360.

Menter, J.M., Golden, J.F. & West, S.S. (1978) Kinetics of fluorescence fading of acridine orange–heparin complexes in solution. *Photochemistry and Photobiology* **27**, 629–633.

Menter, J.M., Hurst, R.E. & West, S.S. (1979) Photochemistry of heparin–acridine orange complexes in solution; photochemical changes occurring in the dye and polymer on fluorescence fading. *Photochemistry and Photobiology* **29**, 473–478.

Miles, R. (1986) Ohmic leakage currents and the dark performance of the 1P21 side-window photomultiplier. *IAPPP Communication* No. 24, 6–21. International Amateur-Professional Photoelectric Photometry, Nashville, TN.

Miller, J.N. (1984) Recent developments in fluorescence and chemiluminescence analysis. *Analyst* **109**, 191–198.

Miller, J.P. & Selverston, A.I. (1979) Rapid killing of single neurons by irradiation of intracellularly injected dye. *Science* **206**, 702–704.

Minsky, M. (1957) Microscopy apparatus. U.S. Patent No. 3013467. Filed 7 November 1957, issued 19 December 1961.

Missmahl, H.P. (1966) Birefringence and dichroism of dyes and their significance in the detection of oriented structures. In: Wied, G.L. (Ed.) *Introduction to quantitative cytochemistry*, pp. 539–547. Academic Press, New York & London.

Mobberley, M.P. (1987) Monitoring the lunar surface with a CCD camera. *Journal of the British Astronomical Association* **97**, 208–210.

Moldavan, A. (1934) Photo-electric technique for the counting of microscopical cells. *Science* **80**, 188–189.

Moneta, G., Brülisauer, M., Jäger, K. & Bollinger, A. (1987) Infrared fluorescence microscopy of skin capillaries with indocyanine green. *International Journal of Microcirculation: Clinical and Experimental* **6**, 25–34.

Moravec, J., Hatt, P.Y., Opie, L.H. & Rost, F.W.D. (1972) The application of the cytophotometer to the study of metabolic transitions of isolated rat heart. *Cardiology* **57**, 61–66.

Moreno, G., Lutz, M. & Bessis, M. (1969) Partial cell irradiation by ultraviolet and visible light: conventional and laser sources. *International Review of Experimental Pathology* **7**, 99–137.

Moreno, G., Salet, C., Kohen, C. & Kohen, E. (1982) Penetration and localization of furocumarins in single living cells studied by microspectrofluorometry. *Biochimica et biophysica acta* **721**, 109–111.

Morgan, C.G. (1987) Time resolved fluorescence microscopy. *Laboratory News* April, pp. 38 and 41.

Moss, D.W. (1960) Kinetics of phosphatase action on naphthyl phosphates, determined by a highly sensitive spectrofluorimetric technique. *Biochemical Journal* **76**, 32P.

Motoda, Y. & Kubota, Y. (1979) Delayed excimer fluorescence of Acridine Orange bound to DNA. *Bulletin of the Chemical Society of Japan* **52**, 693–696.

Mullaney, P.F. & Dean, P.N. (1969) Cell sizing: a small-angle light-scattering method for sizing particles of low refractive index. *Applied Optics* **8**, 2361–2362.

Murray, J.G., Cundall, R.B., Morgan, C.G. & Evans, G.B. (1986) *Journal of Physics E: Scientific Instruments* **19**, 349–355.

Nachlas, M.M., Young, A.C. & Seligman, A.M. (1957) Problems of enzymatic localization by chemical reactions applied to tissue sections. *Journal of Histochemistry and Cytochemistry* **5**, 564–583.

Nairn, R.C., Herzog, F., Ward, H.A. & De Boer, W.G.R.M. (1969) Microphotometry in immunofluorescence. *Clinical and Experimental Immunology* **4**, 697–705.

Nastuk, W.L. (1953) Membrane potential changes at a single muscle end-plate produced by transitory application of acetylcholine with an electrically controlled microjet. *Federation Proceedings* **12**, 102.

Natale, P.J. (1982) Method and reagents for quantitative determination of reticulocytes and platelets in whole blood. US Patent No. 4 336 029.

Neely, J.E., Townend, W.J. & Combs, J.W. (1984) A computerized microspectrophotometer using fibre optics for transmission and detection of light. *Journal of Microscopy* **133**, 313–322.

Nixdorf, J. (1967) Ein neues Verfahren zur Herstellung dünner Drähte mit Durchmessern im um-Bereich. *Drahtwelt* **53**, 696. (Cited in Sernetz, M. & Thaer, A. (1970) A capillary fluorescence standard for microfluorometry. *Journal of Microscopy* **91**, 43–52.)

Nobiling, R. & Bührle, C.P. (1989) A microscope fluorimeter using multiple-wavelength excitation for ultrasensitive single-cell emission spectrometry. *Journal of Microscopy* **156**, 149–161.

Norberg, K.-A., Ritzén, M. & Ungerstedt, U. (1966) Histochemical studies on a special catecholamine-containing cell type in sympathetic ganglia. *Acta physiologica scandinavica* **67**, 260–270.

Nordén, G. (1953) The rate of appearance, metabolism and disappearance of 3,4-benzpyrene in the epithelium of mouse skin after a single application in a volatile solvent. *Acta pathologica et microbiologica scandinavica*, Supplementum **96**. 87 pp.

Nothnagel, E.A. (1987) Quantum counter for correcting fluorescence excitation spectra at 320- to 800-nm wavelengths. *Annals of Biochemistry* **163**, 224–237.

Nuutinen, E.M. (1984) Subcellular origin of the surface fluorescence of reduced nicotinamide nucleotides in the isolated perfused rat heart. *Basic Research in Cardiology* **79**, 49–58.

O'Connor, D.V. & Phillips, D. (1984) *Time-correlated single photon counting*, Chapter 4, pp. 103–131. Academic Press, London.

O'Donnell, C.M. & Solie, T.N. (1976) Fluorometric and phosphorometric analysis. *Analytical Chemistry* **48**, 175–196.

Oldham, P.B., Patonay, G. & Warner, I.M. (1985) Optical output stabilization method for direct current arc lamps. *Review of Scientific Instruments* **56**, 297–302.

Olson, R.A. (1960) Rapid scanning microspectrofluorimeter. *Review of Scientific Instruments* **31**, 844–849.

Olson, R.A., Jennings, W.H. & Butler, W.L. (1964) Molecular orientation: spectral dependence of bifluorescence of chloroplasts *in vivo*. *Biochimica et biophysica acta* **88**, 331–337.

Oostveldt, P.M., Tanke, H.J., Ploem, J.S. & Boeken, G. (1978) Relation between fluorescent intensity and extinction in quantitative microfluorometry. *Acta histochemica* Supplementum **20**, 59–63.

Oster, G. (1955) Birefringence and dichroism. In: Oster, G. & Pollister, A.W. (Eds.) *Physical techniques in biological research*, vol. **1**. Academic Press, New York & London.

Ostling, O. & Johanson, K.J. (1984) Microelectrophoretic study of radiation-induced DNA damages in individual mammalian cells. *Biochemical and Biophysical Research Communications* **123**, 291–298.

Ottenjann, K. (1980) Spektrale Fluoreszenz-Mikrophotometrie von Kohlen und Ölschiefern. *Leitz Mitteilungen für Wissenschaft und Technik* **7**, 262–273. English edition: Spectral fluorescence microphotometry of coal and oil shale, *Leitz Scientific and Technical Information* **7**, 262–273.

Ottenjann, K. (1981/1982) Verbesserungen bei der mikroskopphotometrischen Fluoreszenzmessungen an Kohlenmaceralen. *Zeiss Informationen* (Oberkochen) **26**, 40–46. English edition: Improved microphotometric fluorescence measurements of coal macerals, *Zeiss Information* No. 93E, pp. 40–46.

Ottenjann, K., Teichmüller, M. & Wolf, M. (1974) Spektrale Fluoreszenz-Messungen an Sporiniten mit Auflicht-Anregung, eine mikroskopische Methode zur Bestimmung des Inkohlungsgrades gering inkohlter Kohlen. *Fortschritte in der Geologie von Rheinland und Westfalen* **24**, 2–36.

Ottenjann, K., Teichmüller, M. & Wolf, M. (1975) Spectral fluorescence measurements of sporinites in reflected light and their applicability for coalification studies. In: Alpern, B. (Ed.) *Pétrographie Organique et Potentiel Pétrolier*, pp. 49–65. CNRS, Paris.

Pabst, H. (1980) Application of microspectrophotometry in forensic casework. *Microscopica acta* **S4**, 189–193.

Parker, C.A. (1969a) *Photoluminescence of solutions*. Elsevier, Amsterdam.

Parker, C.A. (1969b) Spectrophosphorimeter microscopy: an extension of fluorescence microscopy. *Analyst* **94**, 161–176.

Parker, C.A. & Rees, W.T. (1960) Correction of fluorescence spectra and measurement of fluorescence quantum efficiency. *Analyst* **85**, 587–600.

Parks, D.R., Hardy, R.R. & Herzenberg, L.L. (1984) Three color immunofluorescence analysis of mouse B lymphocyte subpopulations. *Cytometry* **5**, 159–164.

Parry, W.L. & Hemstreet, G.P. (1988) Cancer detection by quantitative fluorescence image analysis. *Journal of Urology* **139**, 270–274.

Partanen, S. (1978) Carbonyl compound-induced fluorescence of biogenic monoamines in the endocrine cells of the hypophysis. *Progress in Histochemistry and Cytochemistry* **10**, (3). 47 pp.

Partanen, M., Hervonen, A. & Alho, H. (1980) Microspectrofluorimetric estimation of the formaldehyde induced fluorescence of the developing main pelvic ganglion of the rat. *Histochemical Journal* **12**, 49–56.

Partanen, M., Hervonen, A. & Rapaport, S.I. (1982) Microspectrofluorimetric quantitation of histochemically demonstrable catecholamines in peripheral and brain catecholamine-containing neurons in male Fischer-344 rats at different ages. In: Giacobini, E., Filogamo, G., Giacobini, G. & Vernadakis, A. (Eds.) *The aging brain: cellular and molecular mechanisms of aging in nervous system*, vol. **20**, pp. 161–172. Raven Press, New York.

Patzelt, W.J. (1985) *Polarized light microscopy*, 3rd edn. 102 pp. Ernst Leitz GmbH, Wetzlar.

Pavlidis, T. (1982) *Algorithms for graphics image processing*. Computer Science Press, Rockville, MD.

Pearse, A.G.E. (1972) *Histochemistry, theoretical and applied*, 3rd edn, vol. **2**. Churchill Livingstone, London.

Pearse, A.G.E., Ewen, S.W.B. & Polak, J.M. (1972) The genesis of apudamyloid in endocrine polypeptide tumours: histochemical distinction from immunamyloid. *Virchows Archiv B: Zellpathologie* **10**, 93–107.

Pearse, A.G.E., Polak, J.M., Rost, F.W.D., Fontaine, J., Le Lièvre, C. & Le Douarin, N. (1973) Demonstration of the neural crest origin of Type I (APUD) cells in the avian carotid body, using a cytochemical marker system. *Histochemie* **34**, 191–203.

Pearse, A.G.E. & Rost, F.W.D. (1969) A microspectrofluorimeter with epi-illumination and photon counting. *Journal of Microscopy* **89**, 321–328.

Peters, R. (1983) Fluorescence microphotolysis: diffusion measurements in single cells. *Naturwissenschaften* **70**, 294–302.

Peters, R., Brünger, A. & Schulten, K. (1981) Continuous fluorescence microphotolysis: a sensitive method for the study of translational diffusion in single cells. *Proceedings of the National Academy of Sciences, USA* **78**, 962–966.

Peters, R., Peters, J., Tews, K. & Bähr, W. (1974) A microfluorimetric study of translational diffusion in erythrocyte membranes. *Biochimica et biophysica acta* **367**, 282–294.

Petráň, M., Hadravsky, M., Benes, J., Kucera, R. & Boyde, A. (1985) The tandem scanning reflected light microscope. Part I – the principle, and its design. *Proceedings of the Royal Microscopical Society* **20**, 125–129.

Petráň, M., Hadravsky, M. & Boyde, A. (1985) The tandem scanning reflected light microscope. *Scanning* **7**, 97–108.

Petráň, M., Hadravsky, M., Egger, M.D. & Galambos, R. (1968) Tandem scanning reflected light microscope. *Journal of the Optical Society of America* **58**, 661–664.

Phillips, A.P. & Martin, K.L. (1982) Evaluation of a microfluorometer in immunofluorescence assays of individual spores of *Bacillus anthracis* and *Bacillus cereus*. *Journal of Immunological Methods* **49**, 271–282.

Picciolo, G.L. & Kaplan, D.S. (1984) Reduction of fading of fluorescent reaction product for microphotometric quantitation. *Advances in Applied Microbiology* **30**, 197–234.

Pick, J. (1934) Über ein Mikroskop zur Untersuchung lebenden Gewebes. II. Mitteillung. *Zeitschrift für wissenschaftliche Mikroskopie* **51**, 257–262.

Pickard, R. (1986) Getting started in photoelectric photometry. *Journal of the British Astronomical Association* **97**, 14–22.

Piller, H. (1977) *Microscope photometry*. 253 pp. Springer-Verlag, Berlin, Heidelberg & New York.

Pittman, B., Herbert, G.A., Cherry, W.B. & Taylor, G.C. (1967) The quantitation of nonspecific staining as a guide for improvement of fluorescent antibody conjugates. *Journal of Immunology* **98**, 1196–1203.

Plant, A.L., Benson, D.M. & Smith, L.C. (1985) Cellular uptake and intracellular localization of benzo(*a*)pyrene by digital fluorescence imaging microscopy. *Journal of Cell Biology* **100**, 1295–1308.

Platt, J.L. & Michael, A.F. (1983) Retardation of fading and enhancement of intensity of immunofluorescence by *p*-phenylenediamine. *Journal of Histochemistry and Cytochemistry* **31**, 840–842.

Ploeg, M. van der, Broek, K. van den, Smeulders, A.W.M., Vossepoel, A.M. & Duijn, P. van (1977a) HIDACSYS: computer programs for interactive scanning cytophotometry. *Histochemistry* **54**, 273–288.

Ploeg, M. van der & Duijn, P. van (1968) Cytophotometric determination of alkaline phosphatase activity of individual neutrophilic leukocytes with a biochemically calibrated model system. *Journal of Histochemistry and Cytochemistry* **16**, 693–706.

Ploeg, M. van der & Duijndam, W.A.L. (1986) Matrix models. Essential tools for microscopic cytochemical research. *Histochemistry* **84**, 283–300.

Ploeg, M. van der, Vossepoel, A.M., Bosman, F.T & Duijn, P. van (1977b) High-resolution scanning-densitometry of photographic negatives of human chromosomes. *Histochemistry* **51**, 269–291.

Ploem, J.S. (1967) The use of a vertical illuminator with interchangeable dichroic mirrors for fluorescence microscopy with incident light. *Zeitschrift für wissenschaftliche Mikroskopie* **68**, 129–142.

Ploem, J.S. (1969) Ein neuer Illuminator-Typ für die Auflicht-Fluoreszenzmikroskopie. *Leitz-Mitteilungen für Wissenschaft und Technik* **4**, 225–238.

Ploem, J.S. (1970) Standards for fluorescence microscopy. In: Holborow, E.J. (Ed.) *Standardization in immunofluorescence*, pp. 137–153. Blackwell Scientific Publications, Oxford & Edinburgh.

Ploem, J.S. (1977) Quantitative fluorescence microscopy. In: Meek, G.A. & Elder, H.Y. (Eds.) *Analytical and quantitative methods in microscopy*, pp. 55–89. Cambridge University Press, Cambridge.

Ploem, J.S. (1986) New instrumentation for sensitive image analysis of fluorescence in cells and tissues. In: Taylor, D.L., Waggoner, A.S., Lanni, F., Murphy, R.F. & Birge, R.R. (Eds.) *Applications of fluorescence in the biomedical sciences*, pp. 289–300. Alan R. Liss, New York.

Ploem, J.S., de Sterke, J.A., Bonnet, J. & Wasmund, H. (1974) A microspectrofluorometer with epi-illumination operated under computer control. *Journal of Histochemistry and Cytochemistry* **22**, 668–677.

Ploem, J.S. & Tanke, H.J. (1987) *Introduction to fluorescence microscopy*. 56 pp. Oxford University Press/ Royal Microscopical Society, Oxford.

Podgorski, G.T., Longmuir, I.S., Knopp, J.A. & Benson, D.M. (1981) Use of an encapsulated fluorescent probe to measure intracellular P_{O_2}. *Journal of Cellular Physiology* **107**, 329–334.

Poel, J.J. van der, Kardol, M.J., Goulmy, E., Blokland, E., & Bruning, J.W. (1981) Carboxyfluorescein fluorochromasia cell mediated lympholysis. A comparative study. *Immunological Letters* **2**, 187–190.

Pohle, W., Ott, T. & Müller-Welde, P. (1984) Identification of neurons of origin providing the dopaminergic innervation of the hippocampus. *Zeitschrift für Hirnforschung* **25**, 1–10.

Policard, A. & Paillot, A. (1925) Etude de la sécrétion de la soie à l'aide des rayons ultraviolets filtrés (lumière de Wood). *Comptes Rendus de l'Académie des Sciences, Paris* **181**, 378–380.

Popper, H. (1944) Distribution of Vitamin A in tissues visualised by fluorescence microscopy. *Physiological Reviews* **24**, 205–224.

Porter, G. (Ed.) (1967) *Reactivity of the photoexcited organic molecule*. 350 pp. Interscience, New York.

Prenna, G. (1968) Qualitative and quantitative application of fluorescent Schiff-type reagents. *Mikroskopie* **23**, 150–154.

Prenna, G. & Bianchi, U.A. (1964) Reazoni di Feulgen fluorescenti e loro possibilità citofluorometriche quantitative. 4. Studio del processo di fotodecomposizione e confronto fra i dati di estinzione a 470 nm e i data di emissione di fluorescenza nella reactione di Feulgen esquita con Acriflavina-SO$_2$. *Rivista di Istochimica Normale e Patologica* **10**, 645–666.

Prenna, G., Bottiroli, G. & Mazzini, G. (1977) Cytofluorometric quantification of the activity and reaction kinetics of acid phosphatase. *Histochemical Journal* **9**, 15–30.

Prenna, G., Leiva, S. & Mazzini, G. (1974) Quantitation of DNA by cytofluorometry of the conventional Feulgen reaction. *Histochemical Journal* **6**, 467–489.

Prenna, G., Mazzini, G. & Cova, S. (1974) Methodological and instrumental aspects of cytofluorometry. *Histochemical Journal* **6**, 259–278.

Prenna, G. & Sacchi, S. (1964) Primi risultati di microspettrofluorometria qualitativa del tessuto elastico e collagene. *Bollettino della Società Medico-Chirurgica di Pavia* Fascicolo **3–6**, pp. 779–790.

Pringsheim, P. (1963) *Fluorescence and phosphorescence.* John Wiley, New York.

Prosperi, E., Croce, A.C., Bottiroli, G., Dasdia, T. & Supino, R. (1983) Uptake kinetics and intracellular distribution of anthracyclines studied by laser cytofluorometry. *Basic and Applied Histochemistry* **27**, 117–127.

Quaglia, M., Campani, E., Macchiarulo, M., Polacco, E., Columbetti, G., Ghetti, F. & Lenci, F. (1982) A pulsed tunable dye-laser microspectrofluorometer. *Il Nuovo Cimento* **1D**, 382–390.

Querner, F. von (1932) Die paraplasmatischen Einschlusse der Leberzelle im Fluoreszenzmikroskop und der Leuchtstoff X. *Akademische Anzeiger der Akademie der Wissenschaft Wien*, Math.-Naturw. Klasse, No. 18, 7 July.

Raap, A.K. (1986) Localization properties of fluorescence cytochemical enzyme procedures. *Histochemistry* **84**, 317–321.

Redgrave, P. & Mitchell, I. (1982) Photometric assessment of glyoxylic acid-induced fluorescence of dopamine in the caudate nucleus. *Neuroscience* **7**, 871–883.

Reuter, W.O. (1980) Automatisation in microfluorometry. *Microscopica acta* Supplementum **4**, 52–58.

Rich, E.S. & Wampler, J.E. (1981) A flexible, computer-controlled video microscope capable of quantitative spatial, temporal and spectral measurements. *Clinical Chemistry* **27**, 1558–1568.

Rigler, R. (1966) Microfluorometric characterisation of intracellular nucleic acids and nucleoproteins by Acridine Orange. *Acta physiologica scandinavica* **67**, Supplementum **267**. 122 pp.

Ritter, A.W., Tway, P.C., Cline Love, L.J. & Ashworth, H.A. (1981) Microcomputer fluorometer for corrected, derivative and differential spectra and quantum yield determinations. *Analytical Chemistry* **53**, 280–284.

Ritzén, M. (1966a) Quantitative fluorescence microspectrophotometry of catecholamine formaldehyde products. *Experimental Cell Research* **44**, 505–520.

Ritzén, M. (1966b) Quantitative fluorescence microspectrophotometry of 5- hydroxytryptamine-formaldehyde products in models and in mast cells. *Experimental Cell Research* **45**, 178–194.

Ritzén, M. (1967) Cytochemical identification and quantification of biogenic amines. MD thesis, University of Stockholm.

Rosen, S. & Mercer, W.E. (1985) Cytophotometry of breast carcinoma. Acridine-orange DNA microfluorimetry with Giemsa counterstain. *Annals of Quantitative Cytology and Histology* **7**, 159–162.

Rosenfeld, A. & Kak, A.C. (1982) *Digital picture processing*, vols. **1** and **2**, 2nd edn. Academic Press, New York.

Rosselet, A. (1967) Mikrofluorometrische Argininbestimmung. *Zeitschrift für wissenschaftliche Mikroskopie* **68**, 22–41.

Rosselet, A. & Ruch, F. (1968) Cytofluorometric determination of lysine with dansyl chloride. *Journal of Histochemistry and Cytochemistry* **16**, 459–466.

Rost, F.W.D. (1971) Histochemical localization and assay of enzymes. *Journal of Clinical Pathology* **24**, Supplement **4** (Enzyme assays in medicine symposium), 43–50.

Rost, F.W.D. (1972) Cytochemical investigation of substances by fluorescence spectral analysis. PhD thesis, University of London.

Rost, F.W.D. (1973) A microspectrofluorometer for measuring spectra of excitation, emission and absorption in cells and tissues. In: Thaer, A.A. & Sernetz, M. (Eds.) *Fluorescence techniques in cell biology*, pp. 57–63. Springer-Verlag, Berlin, Heidelberg & New York

Rost, F.W.D. (1974) Microspectrofluorometry. *Medical Biology* **52**, 73–81.

Rost, F.W.D. (1980) Quantitative histochemistry. In: Pearse, A.G.E., *Histochemistry, theoretical and applied*, Chapter 11, pp. 379–417. Churchill Livingstone, Edinburgh, London & New York.

Rost, F.W.D. (1991a) *Fluorescence microscopy*, vol. I. Cambridge University Press, Cambridge. (In the press.)

Rost, F.W.D. (1991b) *Fluorescence microscopy*, vol. II. Cambridge University Press, Cambridge. (In the press.)

Rost, F.W.D., Bollmann, R. & Moss, D.W. (1973) Characterization of alkaline phosphatase in tissue sections by microspectrofluorometry. *Histochemical Journal* **5**, 567–575.

Rost, F.W.D. & Ewen, S.W.B. (1971) New methods for the histochemical demonstration of catecholamines, tryptamines, histamine and other arylethylamines by acid- and aldehyde-induced fluorescence. *Histochemical Journal* **3**, 207–212.

Rost, F.W.D., Nägel, L.C.A. & Moss, D.W. (1970) Microfluorimetric investigation of enzyme kinetics in fixed and unfixed tissue sections. *Proceedings of the Royal Microscopical Society* **5**, 76–77.

Rost, F.W.D. & Pearse, A.G.E. (1968) Microspectrophotometry with epi-illumination. *Proceedings of the Royal Microscopical Society* **3**, 22–23.

Rost, F.W.D. & Pearse, A.G.E. (1969) A microspectrofluorimeter with epi-illumination and photon counting. *Journal of Microscopy* **89**, 321–328.

Rost, F.W.D. & Pearse, A.G.E. (1971) An improved microspectrofluorimeter with automatic digital data logging: construction and operation. *Journal of Microscopy* **94**, 93–105.

Rost, F.W.D. & Pearse, A.G.E. (1973) Identification of arylethylamines by microspectrofluorometry of acid- and aldehyde-induced fluorescence. In: Sernetz, M. & Thaer, A.A. (Eds.) *Fluorescence techniques in cell biology*, pp. 199–204. Springer-Verlag, Berlin, Heidelberg & New York.

Rost, F.W.D. & Pearse, A.G.E. (1974) Microfluorometry of primary and secondary fluorescence in biological tissue. *Histochemical Journal* **6**, 245–250.

Rost, F.W.D. & Polak, J.M. (1969) Fluorescence microscopy and microspectrofluorometry of malignant melanomas, naevi and normal melanocytes. *Virchows Archiv A: Pathologische Anatomie* **347**, 321–326.

Rost, F.W.D., Polak, J.M. & Pearse, A.G.E. (1969) The melanocyte: its cytochemical and immunological relationship to cells of the endocrine polypeptide (APUD) series. *Virchows Archiv B: Zellpathologie* **4**, 93–101.

Rost, F.W.D., Polak, J.M. & Pearse, A.G.E. (1973) The cytochemistry of normal and malignant melanocytes, and their relationship to cells of the endocrine polypeptide (APUD) series. *Pigment Cell* **1**, 55–65.

Rotman, B. (1961) Measurement of activity of single molecules of β-D-galactosidase. *Proceedings of the National Academy of Sciences, USA* **47**, 1981–1991.

Rotman, B. & Papermaster, B.W. (1966) Membrane properties of living mammalian cells as studied by enzymatic hydrolysis of fluorogenic esters. *Proceedings of the National Academy of Sciences, USA* **55**, 134–141.

Rousseau, M. (1957) Spectrophotométrie de fluorescence en microscopie. *Bulletin de Microscopie Appliquée* **7**, 92–94.

Rowntree, D. (1981) *Statistics without tears*. 199 pp. Penguin Books, Harmondsworth. Middx.

Ruark, A.E. & Brammer, F.E. (1937) The efficiency of counters and counter circuits. *Physical Review* **52**, 322–324.

Ruch, F. (1957) Dichroismus und Difluoreszenz der Chloroplasten. *Experimental Cell Research Supplement* **4**, 58–62.

Ruch, F. (1960) Ein Mikrospektograph für Absorptionsmessungen im ultravioletten Licht. *Zeitschrift für wissenschaftliche Mikroskopie* **64**, 453–468.

Ruch, F. (1964) Fluoreszenzphotometrie. *Acta histochemica* Supplementum **6**, 117–121.

Ruch, F. (1966a) Birefringence and dichroism of cells and tissue. In: Oster, G. & Pollister, A.W. (Eds.) *Physical techniques in biological research*, vol. **3**, pp. 149–176. Academic Press, New York.

Ruch, F. (1966b) Determination of DNA content by microfluorometry. In: Wied, G.L. (Ed.) *Introduction to quantitative cytochemistry*, pp. 281–294. Academic Press, New York & London.

Ruch, F. (1966c) Dichroism and difluorescence. In: Wied, G.L. (Ed.) *Introduction to quantitative cytochemistry*, pp. 549–555. Academic Press, New York & London.

Ruch, F. (1970) Principles and some applications of cytofluorometry. In: Wied, G. & Bahr, G.F. (Eds.) *Introduction to quantitative cytochemistry II*, pp. 431–454. Academic Press, New York & London.

Ruch, F. (1973) Quantitative determination of DNA and protein in single cells. In: Thaer, A.A. & Sernetz, M. (Eds.) *Fluorescence techniques in cell biology*, pp. 89–93. Springer-Verlag, Berlin, Heidelberg & New York.

Ruch, F. & Bosshard, U. (1963) Photometrische Bestimmung von Stoffmengen im Fluoreszenzmikroskop. *Zeitschrift für wissenschaftliche Mikroskopie* **65**, 335–341.

Ruch, F. & Leemann, U. (1973) Cytofluorometry. In: Neuhoff, V. (Ed.) *Micromethods in molecular biology*, vol. **14**, pp. 331–346. Springer-Verlag, Berlin, Heidelberg & New York.

Ruch, F. & Trapp, L. (1972) A microscope fluorometer with short-time excitation and electronic shutter control. *Zeiss Information* (No. 81) **20**, 59–60.

Rundquist, I. (1981) A flexible system for microscope fluorometry served by a personal computer. *Histochemistry* **70**, 151–159

Rundquist, I. & Enerbäck, L. (1976) Millisecond fading and recovery phenomena in fluorescent biological objects. *Histochemistry* **47**, 79–87.

Rundquist, I. & Enerbäck, L. (1985) A simple microcomputer system for microscope fluorometry. In: Mize, R.D. (Ed.) *The microcomputer in cell and neurobiology research*, pp. 335–353. Elsevier, New York.

Runge, W.J. (1966) A recording microfluorospectrophotometer. *Science* **151**, 1499–1506.

Safranyos, R.G.A., Caveney, S., Miller, J.G. & Petersen, N.O. (1987) Relative roles of gap junction channels and cytoplasm in cell-to-cell diffusion of fluorescent tracers. *Proceedings of the National Academy of Sciences, USA* **84**, 2272–2276.

Sage, B.H., O'Connell, J.P. & Mercolino, T.J. (1983) A rapid vital staining procedure for flow cytometric analysis of human reticulocytes. *Cytometry* **4**, 222–227.

Sahota, T.S., Ibaraki, A., Heywood, F.G., Farris, S.H. & Van Wereld, A. (1981) Image enhancement for light microscopy. *Stain Technology* **56**, 361–366.

Salmon, J.-M., Kohen, E., Viallet, P., Hirschberg, J.G., Wouters, A.W., Kohen, C. & Thorell, B. (1982) Microspectrofluorometric approach to the study of free/bound NAD(P)H ratio as metabolic indicator in various cell types. *Photochemistry and Photobiology* **36**, 585–593.

Salzman, G.C., Growell, J.M. & Martin, J.C. (1975) Cell classification by laser light scattering. Identification and separation of unstained leukocytes. *Acta cytologica* **19**, 374–377.

Scarpelli, D.G. & Pearse, A.G.E. (1958) Physical and chemical protection of cell constituents and the precise localization of enzymes. *Journal of Histochemistry and Cytochemistry* **6**, 369–376.

Schäfer, F.P. (Ed.) (1977) *Dye lasers* (Topics in applied physics, vol. **1**). 2nd edn, 299 pp. Springer-Verlag, Berlin, Heidelberg & New York.

Scheuermann, D.W., Stilman, C. & De Groodt-Lasseel, M.H.A. (1988) Microspectrofluorometric analysis of the formaldehyde-induced fluorophores of 5-hydroxytryptamine and dopamine in intrapulmonary neuroepithelial bodies after administration of L-5-hydroxytryptophan and L-DOPA. *Histochemistry* **88**, 219–226.

Schipper, J. & Tilders, F.J.H. (1979) On the presence of extraneuronal catecholamine in the iris of the rat: scanning microfluorometric study. *Neuroscience Letters* **12**, 229–234.

Schipper, J., Tilders, F.J.H., Groot Wasink, R., Boleij, H.F. & Ploem, J.S. (1980) Microfluorometric scanning of sympathetic nerve fibres: quantification of neuronal and extraneuronal fluorescence with aid of histogram analysis. *Journal of Histochemistry and Cytochemistry* **28**, 124–132.

Schipper, J., Tilders, F.J.H. & Ploem, J.S. (1978) Microfluorimetric scanning of sympathetic nerve fibres: an improved method to quantitate formaldehyde induced fluorescence of biogenic amines. *Journal of Histochemistry and Cytochemistry* **26**, 1057–1066.

Schipper, J., Tilders, F.J.H. & Ploem, J.S. (1979) Extraneuronal catecholamine fluorescence as an index for sympathetic activity: a scanning microfluorimetric study on the iris of the rat. *Journal of Pharmacology and Experimental Therapeutics* **211**, 265–270.

Schipper, J., Tilders, F.J.H. & Ploem, J.S. (1980) A scanning microfluorimetric study on sympathetic nerve fibres: intraneuronal differences in noradrenaline turnover. *Brain Research* **190**, 459–472.

Schlessinger, J., Schechter, Y., Willingham, M.C. & Pastan, I. (1978) Direct visualization of binding, aggregation, and internalization of insulin and epidermal growth factor on living fibroblastic cells. *Proceedings of the National Academy of Sciences, USA* **75**, 2659–2663.

Schneckenburger, H., Feyh, J., Götz, A., Frenz, M. & Brendel, W. (1987) Quantitative *in vivo* measurement of the fluorescent components of Photofrin II. *Photochemistry and Photobiology* **46**, 765–768.

Schneckenburger, H., Pauker, F., Unsöld, E. & Jochum, D. (1985) Intracellular distribution and retention of the fluorescent components of Photofrin II. *Photochemistry and Photobiophysics* **10**, 61–67.

Schneckenburger, H. & Wustrow, T.P.U. (1988) Intracellular fluorescence of photosensitizing porphrins at different concentrations of mitochondria. *Photochemistry and Photobiology* **47**, 471–473.

Schnedl, W., Roscher, U., Ploeg, M. van der & Dann, O. (1977) Cytofluorometric analysis of nuclei and chromosomes by DIPI staining. *Cytobiologie* **15**, 357–362.

Scholz, M., Gross-Johannböcke, C. & Peters, R. (1988) Measurement of nucleo-cytoplasmic transport by fluorescence microphotolysis and laser scanning microscopy. *Cell Biology International Reports* **12**, 709–727.

Scholz, M., Schulten, K. & Peters, R. (1985) Single-cell flux measurement by continuous fluorescence microphotolysis. *European Biophysics Journal* **13**, 37–44.

Seligman, A.M., Chauncey, H.H. & Nachlas, M.M. (1951) Effect of formalin fixation on the activity of five enzymes of rat liver. *Stain Technology* **26**, 19–23.

Sernetz, M. (1973) Microfluorometric investigations on the intracellular turnover of fluorogenic substrates. In: Thaer, A.A. & Sernetz, M. (Eds.) *Fluorescence techniques in cell biology*, pp. 243–254. Springer-Verlag, Berlin, Heidelberg & New York.

Sernetz, M. & Thaer, A. (1970) A capillary fluorescence standard for microfluorometry. *Journal of Microscopy* **91**, 43–52.

Sernetz, M. & Thaer, A. (1973) Microcapillary fluorometry and standardization for microscope fluorometry. In: Thaer, A. & Sernetz, G. (Eds.) *Fluorescence techniques in cell biology*, pp. 41–49. Springer-Verlag, Berlin, Heidelberg & New York.

Serra, J. (1982) *Image analysis and mathematical morphology.* 610 pp. Academic Press, New York.

Serra, J. (1986) Morphological optics. *Journal of Microscopy* **145**, 1–22.

Seul, M. & McConnell, H.M. (1985) Automated Langmuir trough with epifluorescence attachment. *Journal of Physics and Scientific Instruments* **18**, 193–196.

Severin, E. & Stellmach, J. (1984) Impulszytofluorometrie der Redoxactivität von Einzelzellen mit einen neuen fluoreszierenden Formazan. *Acta histochemica* **75**, 101–106.

Severin, E., Stellmach, J. & Nachtigal, H.-M. (1985) Fluorimetric assay of redox activity in cells. *Analytica chimica acta* **170**, 341–346.

Shack, R.V., Bartels, P.H., Buchroeder, R.A., Shoemaker, R.L., Hillman, D.W. & Vukobratovich, D. (1987) Design for a fast fluorescence laser scanning microscope. *Analytical and Quantitative Cytology and Histology* **9**, 509–521.

Shapiro, H.M. (1981) Flow cytometric estimation of DNA and RNA content in intact cells stained with Hoechst 33342 and pyronin Y. *Cytometry* **2**, 143–150.

Shapiro, H.M. (1988) *Practical flow cytometry*, 2nd edn. 353 pp. Alan R. Liss, New York.

Sheppard, C.J.R. (1989) Axial resolution of confocal fluorescence microscopy. *Journal of Microscopy* **154**, 237–241.

Sheppard, C.J.R. & Choudhury (1977) Image formation in the scanning microscope. *Optica acta* **24**, 1051–1073.

Sheppard, C.J.R. & Wilson, T (1978a) Image formation in scanning microscopes with partially coherent source and detector. *Optica acta* **25**, 315–325.

Sheppard, C.J.R. & Wilson, T (1978b) Depth of field in the scanning microscope. *Optics Letters* **3**, 115–117.

Sherman, I.A. & Fisher, M.M. (1986) Hepatic transport of fluorescent molecules: *in vivo* studies using intravital TV microscopy. *Hepatology* **6**, 444–449.

Shoemaker, D.W. & Cummins, J.T. (1976) U.V. laser studies on endogenous brain bioamine-aldehyde condensation products. *Proceedings of the Society of Electro-Optical Instrumentation Engineers* (now *Proceedings of the Society of Photo-Optical Engineers*) **89**, 17–21.

Shotton, D.M. (1988a) Review: video-enhanced light microscopy and its applications in cell biology. *Journal of Cell Science* **89**, 129–150.

Shotton, D.M. (1988b) The current renaissance of light microscopy. II. Blur-free optical sectioning of biological specimens by confocal scanning fluorescence microscopy. *Proceedings of the Royal Microscopical Society* **23**, 289–297.

Sick, T.J. & Rosenthal, M. (1989) Indo-1 measurements of intracellular free calcium in the hippocampal slice: complications of labile NADH fluorescence. *Journal of Neuroscience Methods* **28**, 125–132.

Sidgwick, J.B. (1979) *Amateur astronomer's handbook*, 4th edn, ed. Muirden, J. 568 pp. Pelham Books, London.

Sims, P.J., Waggoner, A.S., Wang, C.-M. & Hoffmann, J.F. (1974) Studies on the mechanism by which cyanine dyes measure membrane potential in red blood cells and phosphatidyl-choline vesicles. *Biochemistry* **13**, 3315–3330.

Singer, E. (1932) A microscope for observations of fluorescence in living tissues. *Science* **75**, 289–291.

Sisken, J.E., Barrows, G. & Grasch, S.D. (1983) Quantitation of fluorescence at the cellular level with a new computer-based video photoanalysis system. *Journal of Cell Biology* **95**, 462a.

Sisken, J.E., Barrows, G.H. & Grasch, S.D. (1986) The study of fluorescent probes by quantitative video intensification microscopy (QVIM). *Journal of Histochemistry and Cytochemistry* **34**, 61–66.

Sisken, J.E., Silver, R.B., Barrows, G.H. & Grasch, S.D. (1985) Studies on the role of Ca^{++} in cell division with the use of fluorescent probes and quantitative video intensification microscopy. *Progress in Clinical and Biological Research* **196**, 73–87.

Slayter, E.M. (1970) *Optical methods in biology*. 757 pp. Wiley-Interscience, New York.

Slomba, A.F., Wasserman, D.E., Kaufman, G.I. & Nester, J.F. (1972) A laser flying spot scanner for use in automated fluorescence antibody instrumentation. *Journal of the Association of Advanced Medical Instrumentation* **6**, 230–234.

Smith, B.A., Clark, W.R. & McConnell, H.M. (1979) Anisotropic molecular motion on cell surfaces. *Proceedings of the National Academy of Sciences, USA* **76**, 5641–5644.

Smith, B.A. & McConnell, H.M. (1978) Determination of molecular motion in membranes using periodic pattern photobleaching. *Proceedings of the National Academy of Sciences, USA* **75**, 2759–2763.

Smith, L.M., McConnell, H.M, Smith, B.A & Parce, J.W. (1981) Pattern photobleaching of fluorescent lipid vesicles using polarized laser light. *Biophysical Journal* **33**, 139–146.

Smith, M.T., Redick, J.A. & Baron, J. (1983) Quantitative immunohistochemistry: a comparison of microdensitometric analysis of unlabeled antibody peroxidase–antiperoxidase staining and of microfluorometric analysis of indirect fluorescent antibody staining for nicotinamide adenosine dinucleotide phosphate (NADPH)–cytochrome *c* (P-450) reductase in rat liver. *Journal of Histochemistry and Cytochemistry* **31**, 1183–1189.

Snowdon, L.R., Brooks, P.W. & Goodarzi, F. (1986) Chemical and petrological properties of some liptinite-rich coals from British Columbia. *Fuel* **65**, 459–472.

Soini, E. & Kojala, H. (1983) Time-resolved fluorometer for lanthanide chelates: a new generation of nonisotopic immunoassays. *Clinical Chemistry* **29**, 65–68.

Spackman, W., Davis, A. & Mitchell, G.D. (1976) The fluorescence of liptinite macerals. *Brigham Young University Geological Studies* **22**, 59–91.

Spatz, W.B. & Grabig, S. (1983) Reduced facing of fast blue fluorescence in the brain of the guinea-pig by treatment with sodium-nitroprusside. *Neuroscience Letters* **38**, 1–4.

Sprenger, E. & Böhm, N. (1971a) Qualitative und quantitative Fluoreszenzmikrospekrographie mit dem LEITZ-Mikrospektrographen. *Histochemie* **25**, 163–176.

Sprenger, E. & Böhm, N. (1971b) Der Einfluss von Metachromasie und Photodecomposition auf die quantitative Feulgen-DNS Fluoreszenzcytophotometrie. *Histochemie* **25**, 171–181.

Spring, K.R. & Smith, P.D. (1987) Illumination and detection systems for quantitative fluorescence microscopy. *Journal of Microscopy* **147**, 265–278.

Stach, E. (1982) Fluorescence microscopy. In: Stach, E., Mackowsky, M.-T., Teichmüller, M., Taylor, G.H., Chandra, D. & Teichmüller, R. *Stach's textbook of coal petrology*, 3rd edn, pp. 348–356. Gebruder Borntraeger, Stuttgart.

Stach, E., Mackowsky, M.-T., Teichmüller, M., Taylor, G.H., Chandra, D. & Teichmüller, R. (1975) *Stach's textbook of coal petrology*, 2nd edn, pp. 5–54. Gebruder Borntraeger, Stuttgart.

Stair, R., Johnston, R.G. & Halbach, E.W. (1960) Standard of spectral radiance for the region of 0·25 to 2·6 microns. *Journal of Research of the National Bureau of Standards – A. Physics and Chemistry* **64A**, 291–296.

Stair, R., Schneider, W.E. & Jackson, J.K. (1963) A new standard of spectral irradiance. *Applied Optics* **2**, 1151–1154.

Steel, R.G.D. & Torrie, J.H. (1980) *Principles and procedures of statistics: a biomedical approach*, 2nd edn. 633 pp. McGraw-Hill, New York.

Steen, H.B. (1980) Further developments of a microscope-based flow cytometry: light scatter detection and excitation compensation. *Cytometry* **1**, 26–31.

Steinkamp, J. & Stewart, C. (1986) Dual-laser differential fluorescence correction method for reducing cellular background autofluorescence. *Cytometry* **1**, 566–574.

Stellmach, J. (1984) Fluorescent redox dyes. 1. Production of fluorescent formazan by unstimulated and phorbol ester- or digitonin-stimulated Ehrlich ascites tumor cells. *Histochemistry* **80**, 137–143.

Stepanov, B.I. (1957) A universal relation between the absorption and luminescence spectra of complex molecules. *Doklady Akademii Nauk SSR* **112**, 839–842. English translation: *Soviet Physics Doklady* **2**, 81–84.

Steponkus, P.L., Dowgert, M.F., Ferguson, J.R. & Levin, R.L. (1984) Cryomicroscopy of isolated plant protoplasts. *Cryobiology* **21**, 209–233.

Steppel, R. (1982) Organic dye lasers. In: Wever, M.J. (Ed.) *Handbook of science and technology*, vol. **1**. CRC Press, Boca Raton, FL.

Sternberger, L.A. (1986) *Immunocytochemistry*, 3rd edn. 524 pp. John Wiley & Sons, New York.

Stoehr, M., Eipel, H. & Goerttler, K. (1977) Extended applications of flow microfluorometry by means of dual laser excitation. *Histochemistry* **51**, 305–313.

Stoehr, M. & Futterman, G. (1979) Visualization of multidimensional spectra in flow cytometry. *Journal of Histochemistry and Cytochemistry* **27**, 560–563.

Stoehr, M., Vogt-Schaden, M., Knobloch, M., Vogel, R. & Futterman, G. (1978) Evaluation of eight fluorochrome combinations for simultaneous DNA-protein flow analysis. *Stain Technology* **53**, 205–215.

Stoltz, J.F. & Donner, M. (1985) Fluorescence polarization applied to cellular microrheology. *Biorheology* **22**, 227–247.

Storz, H. & Jelke, E. (1984) Photomicrography of weakly fluorescent objects – employment of paraphenylene diamine as a blocker of fading. *Acta histochemica* **75**, 133–140.

Stoward, P.J. (1968a) A simple microfluorometer. *Proceedings of the Royal Microscopical Society* **3**, 122–125.

Stoward, P.J. (1968b) Studies in fluorescence histochemistry. VI. Fluorescence fading rates of mucosubstance salicylhydrazones and their aluminium complexes. *Journal of the Royal Microscopical Society* **88**, 587–593.

Stoward, P.J. (1980) Criteria for the validation of quantitative histochemical enzyme techniques. In: *Trends in enzyme histochemistry and cytochemistry*, Ciba Foundation Symposium **73**, 11–31.

Takamatsu, T. & Fujita, S. (1988) Microscopic tomography by laser scanning microscopy and its three-dimensional reconstruction. *Journal of Microscopy* **149**, 167–174.

Takamatsu, T., Kitamura, T. & Fujita, S. (1986) Quantitative fluorescence image analysis. *Acta histochemica cytochemica* **19**, 61–71.

Täljedal, I.B. (1970) Direct fluorophotometric recording of enzyme kinetics in cryostat sections. *Histochemie* **21**, 307–313.

Talmi, Y. (Ed.) (1983) *Multichannel image detectors*, vol. **2**. ACS Symposium Series, No. 236. American Chemical Society, Washington, DC.

Tanke, H.J. (1989) Does light microscopy have a future? *Journal of Microscopy* **155**, 405–418.

Tanke, H.J., Deelder, A.M., Dresden, M.H., Jongkind, J.F. & Ploem, J.S. (1985) The aperture-defined microvolume (ADM) method: automated measurements of enzyme activity using an inverted fluorescence microscope. *Histochemical Journal* **17**, 797–804.

Tanke, H.J. & Ingen, E.M. van (1980) A reliable Feulgen-acriflavine-SO$_2$ staining procedure for quantitative DNA measurements. *Journal of Histochemistry and Cytochemistry* **28**, 1007–1013.

Tanke, H.J., Nieuwenhuis, I.A.B., Koper, G.J.M., Slats, J.C.M. & Ploem, J.S. (1980) Flow cytometry of human reticulocytes based on RNA fluorescence. *Cytometry* **1**, 313–320.

Tanke, H.J., Oostveldt, P. van & Duijn, P. van (1982) A parameter for the distribution of fluorophores in cells derived from measurements of inner filter effect and reabsorption phenomenon. *Cytometry* **2**, 359–369.

Tanke, H.J., Rothbarth, P.H., Vossen, J.M.J.J., Koper, G.J.M. & Ploem, J.S. (1983) Flow cytometry of reticulocytes applied to clinical haematology. *Blood* **61**, 1091–1097.

Tanke, H.J., Deelder, A.M., Dresden, M.H., Jongkind, J.F. & Ploem, J.S. (1985) The aperture-defined microvolume (ADM) method: automated measurements of enzyme activity using an inverted fluorescence microscope. *Histochemical Journal* **17**, 797–804.

Tanke, H.J., Vianen, P.H. van, Emiliani, F.M.F., Neuteboom, I., de Vogel, N., Tates, A.D., de Bruijn, E.A. & Oosterom, A.T. van (1986) Changes in erythropoiesis due to radiation or chemotherapy as studied by flow cytometric determination of peripheral blood reticulocytes. *Histochemistry* **84**, 544–548.

Taylor, C.E.D. & Heimer, G.V. (1975) Quantitative immunofluorescence studies. In: Hijmans, W. & Schaeffer, M. (Eds.) Fifth international conference on immunofluorescence and related staining techniques. *Annals of the New York Academy of Sciences* **254**, 151–156.

Taylor, C.E.D., Heimer, G.V. & Lidwell, O.M. (1971) Use of a fibre optic probe for quantitative immunofluorescence. *Lancet* **i**, 785–786.

Taylor, D.L., Amato, P.A., McNeil, P.L., Luby-Phelps, K. & Tanasugarn, L. (1986) Spatial and temporal dynamics of specific molecules and ions in living cells. In: Taylor, D.L, Waggoner, A.S., Lanni, F., Murphy, R.F. & Birge, R.R. (Eds.) *Applications of fluorescence in the biomedical sciences*, pp. 347–376. Alan R. Liss, New York.

Teerman, S.C., Crelling, J.C. & Glass, G.B. (1987) Fluorescence spectral analysis of resinite macerals from coals of the Hanna Formation, Wyoming, U.S.A. *International Journal of Coal Geology* **7**, 315–334.

Teichmüller, M. & Durand, B. (1983) Fluorescence microscopical rank studies on liptinites and vitrinites in peat and coals, and comparison with results of the rock-eval pyrolysis. *International Journal of Coal Geology* **2**, 197–230.

Teichmüller, M. & Ottenjann, K. (1977) Liptinite and lipoide Stoffe in einem Erdölmuttergestein. Art und Diagenese von Liptiniten und lipoiden Stoffen in einem Erdölmuttergestein aufgrund fluoreszenzmikroskopischer Untersuchungen. *Erdöl und Kohle* **30**, 387–398.

Teichmüller, M. & Wolf, M. (1977) Application of fluorescence microscopy in coal petrology and oil exploration. *Journal of Microscopy* **109**, 49–73.

Thaer, A. (1966a) Instrumentation for microfluorometry. In: Wied, G. (Ed.) *Introduction to quantitative cytochemistry*, pp. 409–426. Academic Press, New York.

Thaer, A. (1966b) Speicherbetrieb mit einem Superorthicon in der Mikrospektrographie. Haus der Technik-Vortragsveroffentlichungen, Heft 69/66. *Fernsehen in der Industrie* 49–55.

Thaer, A. & Becker, H. (1975) Microscope fluorometric investigations on the reticulocyte maturation distribution as diagnostic criterion of disordered erythopoiesis. *Blut* **30**, 339–348.

Thaer, A.A. & Sernetz, M. (1973) *Fluorescence techniques in cell biology*. 420 pp. Springer-Verlag, Berlin, Heidelberg & New York.

Thieme, G.A. (1966) A versatile device for microscopic spectrofluorometry. *Acta physiologica scandinavica* **67**, 514–520.

Thomas, J.A., Buschbaum, R.N., Zimmick, A. & Racker, E. (1979) Intracellular pH measurements in Ehrlich ascites tumour cells utilizing spectroscopic probes generated *in situ*. *Biochemistry* **18**, 2210–2218.

Thornhill, D.P. (1975) Separation of a series of chromophores and fluorophores present in elastin. *Biochemical Journal* **147**, 215–219.

Thyagarajan, K. & Ghatak, A.K. (1981) *Lasers, theory and applications.* 431 pp. Plenum Press, New York.

Tiffe, H.-W. (1975) Microfluorometric measurements at low temperature. *Histochemistry* **45**, 77–81.

Tiffe, H.-W. (1977) A microscope stage sample holder for microfluorometric measurements of biological specimen in the range of temperature between 3·5 to 300 K. *Histochemistry* **52**, 171–177.

Tiffe, H.-W. & Hundeshagen, M. (1982) Investigation of fading and recovery of fluorescence intensity at 73·5 K. *Journal of Microscopy* **126**, 231–235.

Tilders, F.J.H., Ploem, J.S. & Smelik, P.G. (1974) Quantitative microfluorometric studies on formaldehyde-induced fluorescence of 5-hydroxytryptamine in the pineal gland of the rat. *Journal of Histochemistry and Cytochemistry* **22**, 967–975.

Traganos, F., Darzynkiewicz, Z., Sharpless, T. & Melamed, M.R. (1977) Simultaneous staining of ribonucleic and deoxyribonucleic acids in unfixed cells using acridine orange in a flow cytofluorometric system. *Journal of Histochemistry and Cytochemistry* **25**, 46–56.

Trask, B., Engh, G. van den, Landegent, J., Jansen in de Wal, N. & Ploeg, M. van der (1985) Detection of DNA sequences in nuclei in suspension by in situ hybridization and dual beam flow cytometry. *Science* **20**, 1401–1403.

Tsien, R.Y. (1980) New calcium indicators and buffers with high selectivity against magnesium and protons: design, synthesis and properties of prototype structures. *Biochemistry* **19**, 2396–2404.

Tsien, R.Y. (1988) Fluorescent measurement and photochemical manipulation of cytosolic free calcium. *Trends in Neurosciences* **11**, 419–424.

Tsien, R.Y., Pozzan, T. & Rink, T.J. (1984) Measuring and manipulating cytosolic Ca^2 + with trapped indicators. *Trends in Biochemical Science* **9**, 263–266.

Tsuchihashi, Y., Nakanishi, K., Fukuda, M. & Fujita, S. (1979) Quantification of nuclear DNA and intracellular glycogen in a single cell by fluorescent double-staining. *Histochemistry* **63**, 311–322.

Udenfriend, S. (1962) *Fluorescence assay in biology and medicine.* Academic Press, New York & London.

Udenfriend, S. (1969) *Fluorescence assay in biology and medicine*, 2nd edn. Academic Press, New York.

Uematsu, D., Greenberg, J.H., Reivich, H., Kobayashi, S. & Karp, A. (1988) In vivo fluorometric measurement of changes in cytosolic free calcium from the cat cortex during anoxia. *Journal of Cerebral Blood Flow and Metabolism* **8**, 367–374.

Valet, G., Raffael, A., Moroden, L., Wursch, E. & Ruhenstroth-Bauer, G. (1981) Fast intracellular pH determination in single cells by flow cytometry. *Naturwissenschaften* **68**, 265–266.

Valnes, K. & Brandtzaeg, P. (1985) Retardation of immunofluorescence fading during microscopy. *Journal of Histochemistry and Cytochemistry* **33**, 755–761.

van, van der – for names beginning with van or van der, see under the capitalized surname.

Van Dilla, M.A., Dean, P.N., Laerum, O.D. & Melamed, M.R. (1985) *Flow cytometry: instrumentation and data analysis.* Series on Analytical Cytology, ed. Ploem, J.S. Academic Press, London.

Van Dilla, M.A., Trujillo, T.T., Mullaney, P.F. & Coulter, J.R. (1969) Cell microfluorometry: a method for rapid fluorescence measurements. *Science* **163**, 1212–1214.

Van Orden, L.S. (1970) Quantitative histochemistry of biogenic amines. A simple microspectrofluorometer. *Biochemical Pharmacology* **19**, 1105–1117.

Van Orden, L.S., Vugman, I., Bensch, K.G. & Giarman, N.J. (1967) Biochemical, histochemical and electron-microscopic studies of 5-hydroxytrypamine in neoplastic mast cells. *Journal of Pharmacology and Experimental Therapeutics* **158**, 195–205.

Van Orden, L.S., Vugman, I. & Giarman, N.J. (1965) 5-Hydroxytryptamine in single neoplastic mast cells. A microscopic spectrofluorometric study. *Science* **148**, 162–164.

Vassy, J., Rigaut, J.P., Hill, A.-M. & Foucrier, J. (1990) Analysis by confocal scanning laser microscopy imaging of the spatial distribution of intermediate filaments in foetal and adult rat liver cells. *Journal of Microscopy* **157**, 91–104.

Vaughan, A., Guilbault, G.C. & Hackney, D. (1971) Fluorometric methods for analysis of acid and alkaline phosphatase. *Analytical Chemistry* **43**, 721–724.

Vaughan, W.M. & Weber, G. (1970) Oxygen quenching of pyrenebutyric acid fluorescence in water. A dynamic probe of the microenvironment. *Biochemistry* **9**, 464–473.

Vialli, M. & Prenna, G. (1969) Contribution to the cytospectrofluorometric measurement of 5-

hydroxytryptamine in enterochromaffin cells. *Journal of Histochemistry and Cytochemistry* **15**, 321–330.

Vindelov, L.L., Christensen, I.J. & Nissen, N.I. (1983) A detergent-trypsin method for the preparation of nuclei for flow cytometric DNA analysis. *Cytometry* **3**, 323–327.

Visser, J.W.M., Jongling, A.A.M. & Tanke, H.J. (1979) Intracellular pH-determination by fluorescence measurements. *Journal of Histochemistry and Cytochemistry* **27**, 32–35.

von – for names beginning with von, see under the capitalized surname.

Vrolijk, J., Tenbrinke, H., Ploem, J.S. & Pearson, P.L. (1980) Video techniques applied to chromosome analysis. *Microscopica acta* **S4**, 108–115.

Wade, M.H., Trosko, J.E. & Schindler, M. (1986) A fluorescence photobleaching assay of gap junction-mediated communication between human cells. *Science* **232**, 429–552.

Waggoner, A.S. (1985) Dye probes of cell organelle and vesicle membrane potentials. In: Martinosi, A. (Ed.) *The enzymes of biological membranes*, pp. 313–331. Plenum Press, New York.

Waggoner, A.S. (1986) Fluorescent probes for analysis of cell structure, function and health by flow and imaging cytometry. In: Taylor, D.L., Waggoner, A.S., Lanni, F., Murphy, R.F. & Birge, R.R. (Eds.) *Applications of fluorescence in the biomedical sciences*, pp. 3–28. Alan R. Liss, New York.

Wagner, S., Feldman, A. & Snipes, W. (1982) Recovery from damage induced by acridine plus near-ultraviolet light in *Escherichia coli. Photochemistry and Photobiology* **35**, 73–81.

Wahl, F.M. (1987) *Digital image signal processing.* 190 pp. Artech House Books, Boston.

Wahl, P. (1975) Nanosecond pulse fluorometry. In: Pain, R. & Smith, B. (Eds.) *New techniques in biophysics and cell biology*, pp. 233–285. John Wiley & Sons, Chichester.

Wahren, B. (1978) Cellular content of carcinoembryonic antigen in urothelial carcinoma. *Cancer* **42** (3 Suppl.), 1533–1539.

Walpole, R.E. & Myers, R.H. (1978) *Probability and statistics for engineers and scientists*, 2nd edn. 580 pp. Collier Macmillan, New York.

Walter, R.J. & Berns, M.W. (1981) Computer-enhanced video microscopy: digitally processed microscope images can be produced in real-time. *Proceedings of the National Academy of Sciences, USA* **78**, 6927–6931.

Ware, B.R., Brvenik, L.J., Cummings, R.T., Furukawa, R.H. & Krafft, G.A. (1986) Fluorescence photoactivation and dissipation (FPD). In: Taylor, D.L., Waggoner, A.S., Murphy, R.F., Lanni, F. & Birge, R.R. (Eds.) *Applications of fluorescence in the biomedical sciences*, pp. 141–157. Alan R. Liss, New York.

Ware, W.R. & Baldwin, B.A. (1964) Absorption intensity and fluorescence lifetimes of molecules. *Journal of Chemical Physics* **40**, 1703–1705.

Wasmund, H. & Nickel, W. (1973) Ein neues universelles Steuergerät für mikrophotometrische Untersuchungsmethoden. *Leitz Mitteilungen für Wissenschaft und Technik* **6**, 71–72.

Watkinson, J. (1988) *The art of digital audio.* 489 pp. Focal Press, London.

Watson, J.V. (1987) Quantitation of molecular and cellular probes in populations of single cells using fluorescence. *Molecular and Cellular Probes* **1**, 121–136.

Watson, J.V., Sikora, K.E. & Evan, G.I. (1985) A simultaneous flow cytometric assay for c-myc oncoprotein and cellular DNA in nuclei from paraffin-embedded material. *Journal of Immunological Methods* **83**, 179–192

Webb, W.W. & Gross, D. (1986) Patterns of individual molecular motions deduced from fluorescent image analysis. In: Taylor, D.L., Waggoner, A.S., Lanni, F., Murphy, R.F. & Birge, R.R. (Eds.) *Applications of fluorescence in the biomedical sciences*, pp. 405–422. Alan R. Liss, New York.

Weber, G. (1961) Enumeration of components in complex systems by fluorescence spectrophotometry. *Nature* **190**, 27–29.

Weber, G. & Teale, F.W.J. (1957) Fluorescence excitation spectrum of organic compounds in solution. *Transactions of the Faraday Society* **54**, 640–648.

Weber, K. (1965a) Leitz-Mikroskop-Photometer MPV mit variabler Messblende. *Leitz-Mitteilungen für Wissenschaft und Technik* **3**, 103–107.

Weber, K. (1965b) Strahlungsquellen für die Mikrophotometrie. *Acta histochemica (Jena)* Supplementum **6**, 157–163.

Wehland, J. & Weber, K. (1980) Distribution of fluorescently labelled actin and tropomyosin after microinjection in living tissue culture cells as observed with TV image intensification. *Experimental Cell Research* **127**, 397–408.

Wehry, E.L. (1973) Effects of molecular structure and molecular environment on fluorescence. In: Guilbault, G.G. (1973) (Ed.) *Practical fluorescence: theory, methods and techniques*, Chapter 3, pp. 79–136. Marcel Dekker, New York.

Wehry, E.L. (1982) Molecular fluorescence, phosphorescence and chemiluminescence spectrometry. *Analytical Chemistry Review* **54**, 131–150.

Weiss, M.J. & Chen, L.B. (1984) Rhodamine 123: a lipophilic mitochondrial-specific vital dye. *Kodak Laboratory Chemical Bulletin* **55**, 1–4.

Weissenböck, G., Schnabl, H., Scharf, H. & Sachs, G. (1987) On the properties of fluorescing compounds in guard and epidermal cells of *Allium cepa* L. *Planta* **171**, 88–95.

Werner, I.M., Christian, G.D., Davidson, E.R. & Callis, J.B. (1977) Analysis of multicomponent fluorescence data. *Analytical Chemistry* **49**, 564–573.

West, S.S. (1965) Fluorescence microscopy of mouse leucocytes supravitally stained with Acridine Orange. *Acta histochemica* Supplementum **6**, 135–153.

West, S.S. & Golden, J.F. (1976) Phosphor particles as microscope fluorescence standards. *Journal of Histochemistry and Cytochemistry* **24**, 609–610.

West, S.S., Loeser, C.N. & Schoenberger, M.D. (1960) Television spectroscopy of biological fluorescence. *Institute of Radio Engineers Transactions of Medical Electronics* ME-7, 138–142.

White, J.G., Amos, W.B. & Fordham, M. (1987) An evaluation of confocal versus conventional imaging of biological structures by fluorescence light microscopy. *Journal of Cell Biology* **105**, 41–48.

White, C.E., Ho, M. & Weimer, E.Q. (1960) Methods for obtaining correction factors for fluorescence spectra as determined with the Aminco-Bowman spectrophotofluorometer. *Analytical Chemistry* **32**, 438–440.

Wick, G., Baudner, S. & Herzog, F. (1978) *Immunofluorescence*. Die Medizinische Verlagsgesellschaft, Marburg/Lahn.

Wick, G., Schauenstein, K., Herzog, F. & Stainbatz, A. (1975) Investigations of the recovery phenomenon after laser excitation in immunofluorescence. In: Hijmans, W. & Schaeffer, M. (Eds.) Fifth international conference on immunofluorescence and related staining techniques. *Annals of the New York Academy of Sciences* **254**, 151–156.

Widholm, J.M. (1972) Use of fluorescein diacetate and phenosafranine for determining viability of cultured plant cells. *Stain Technology* **47**, 189–194.

Wied, G.L. (Ed.) (1966) *Introduction to quantitative cytochemistry*. 623 pp. Academic Press, New York & London.

Wied, G.L. & Bahr, G.F. (Eds.) (1970) *Introduction to quantitative cytochemistry II*. 551 pp. Academic Press, New York & London.

Wijnaendts van Resandt, R.W., Marsman, H.J.B., Kaplan, R., Davoust, J., Stelzer, E.H.K. & Stricker, R. (1985) Optical fluorescence microscopy in three dimensions: microtomoscopy. *Journal of Microscopy* **138**, 29–34.

Wilke, V. (1985) Scanning optical microscopy – the Laser Scan microscoope. *Scanning* **7**, 88–95.

Wilkinson, G.N. (1961) Statistical estimations in enzyme kinetics. *Biochemical Journal* **80**, 324–332.

Williams, D.A., Fogarty, K.E., Tsien, R.Y. & Fay, F.S. (1985) Calcium gradients in single smooth muscle cells revealed by the digital imaging microscopy using Fura-2. *Nature* **318**, 558–561.

Williamson, J.R. & Jamieson, D. (1965) Dissociation of the inotropic from the glycogenolytic effect of epinephrine in the isolated rat heart. *Nature* **206**, 364–367.

Willingham, M.C. & Pastan, I. (1978) The visualization of fluorescent proteins in living cells by video intensification microscopy (VIM). *Cell* **13**, 501–507.

Wilson, T. (1989) Optical sectioning in confocal fluorescent microscopes. *Journal of Microscopy* **154**, 143–156.

Wilson, T. & Carlini, A.R. (1987) Size of the detector in confocal imaging systems. *Optics Letters* **12**, 227–229.

Wilson, T. & Carlini, A.R. (1989) Aberrations in confocal imaging systems. *Journal of Microscopy* **154**, 243–256.

Wilson, T. & Sheppard, C.J.R. (1984) *Theory and practice of scanning optical microscopy.* 213 pp. Academic Press, London.

Wittig, B., Rohrer, F. & Zetzsch, C. (1984) Intelligent microcomputer interface for continuous registration and storage of spectra by photon counting. *Review of Scientific Instruments* **55**, 375–378.

Wotherspoon, N., Oster, G.K. & Oster, G. (1972) The determination of fluorescence and phosphorescence. In: Weisberger, A. & Rossiter, B.W. (Eds.) *Physical methods of organic chemistry.* Wiley-Interscience, New York.

Wouters, C.H., Gevel, J.S. van der, Meer, J.W.M. van der, Daems, W.T., Furth, R. van & Ploem, J.S. (1987) Cell surface characteristics and DNA content of macrophages in murine bone marrow cells. *Histochemistry* **86**, 433–436.

Wreford, N.G.M. & Schofield, G.C. (1975) A microspectrofluorometer with on line real time correction of spectra. *Journal of Microscopy* **103**, 127–130.

Wreford, N.G. & Smith, G.C. (1979) Differentiation of the formaldehyde-induced fluorescent products of noradrenalin and dopamine by microspectrofluorometry. *Histochemical Journal* **11**, 473–483.

Wreford, N.G.M. & Smith, G.C. (1982) Microspectrofluorometry in biogenic amine research. *Brain Research Bulletin* **9**, 87–96.

Yamada, M., Takaksu, A., Yamamoto, K. & Iwata, S. (1966) Microfluorometry of nucleic acids in the cell involving Caspersson–Riglers method. *Archivum histologicum japonicum* **27**, 387–395.

Yanagida, M., Hiraoka, Y., Matsumoto, S., Uemura, T. & Okada, S. (1986) Spatial and temporal dynamics of specific molecules and ions in living cells. In: Taylor, D.L., Waggoner, A.S., Lanni, F., Murphy, R.F. & Birge, R.R. (Eds.) *Applications of fluorescence in the biomedical sciences,* pp. 321–346. Alan R. Liss, New York.

Yanagida, M., Morikawa, K., Hiraoka, Y. & Katsura, I. (1983) Dynamic behaviors of DNA molecules in solution studied by fluorescence microscopy. *Cold Spring Harbour Symposia on Quantitative Biology* **47**, 177–187.

Zalewski, E.F., Geist, J. & Velapoldi, R.A. (1982) Correcting emission and excitation spectra: a review of past procedures and new possibilities using silicon photodiodes. In: DeLyle Eastwood (Ed.) *New directions in molecular luminescence. ASTM Special Technical Publication* **822** pp. 103–111.

Ziegenspeck, H. (1949) Die Emission polarisierten Fluoreszenzlichtes (Difluoreszenz) durch gefärbte Zellulose und Kutinmembranen von Pflanzen. In: Bräutigam, F. & Grabner, A. (Eds.) *Beitrage zur Fluoreszenzmikroskopie* (a special volume of *Mikroskopie*), pp. 71–85. Verlag Georg Fromme & Co., Wien.

Zirkle, R.E. (1957) Partial cell irradiation. *Advances in Biological and Medical Physics* **5**, 103–146.

Zs.-Nagy, I., Ohta, M. & Kitani, K. (1989) Effect of centrophenoxine as BCE-001 treatment on lateral diffusion of proteins in the hepatocyte plasma membrane as revealed by fluorescence recovery after photobleaching in rat liver smears. *Experimental Gerontology* **24**, 317–330.

Zs.-Nagy, I., Ohta, M., Kitani, K. & Imahori, K. (1984) An automated method for measuring lateral motility of proteins in the plasma membrane of cells in compact tissues by means of fluorescence recovery after photobleaching. *Mikroskopie* (Wien) **41**, 12–25.

Zweig, A. (1973) Photochemical generation of stable fluorescent compounds (photofluorescence). *Pure and Applied Chemistry* **33**, 389–410.

Index

Page numbers in *italics* refer only to tables or figures.